Medical Response Strategy in Case of Radiation Emergency Caused by the War in Ukraine

NATO Science for Peace and Security Series

This Series presents the results of scientific meetings supported under the NATO Programme: Science for Peace and Security (SPS).

The NATO SPS Programme supports meetings in the following Key Priority areas: (1) Defence Against Terrorism; (2) Countering other Threats to Security and (3) NATO, Partner and Mediterranean Dialogue Country Priorities. The types of meeting supported are generally "Advanced Study Institutes" and "Advanced Research Workshops". The NATO SPS Series collects together the results of these meetings. The meetings are coorganized by scientists from NATO countries and scientists from NATO's "Partner" or "Mediterranean Dialogue" countries. The observations and recommendations made at the meetings, as well as the contents of the volumes in the Series, reflect those of participants and contributors only; they should not necessarily be regarded as reflecting NATO views or policy.

Advanced Study Institutes (ASI) are high-level tutorial courses intended to convey the latest developments in a subject to an advanced-level audience

Advanced Research Workshops (ARW) are expert meetings where an intense but informal exchange of views at the frontiers of a subject aims at identifying directions for future action

Following a transformation of the programme in 2006 the Series has been re-named and re-organised. Recent volumes on topics not related to security, which result from meetings supported under the programme earlier, may be found in the NATO Science Series.

The Series is published by IOS Press, Amsterdam, and Springer, Dordrecht, in conjunction with the NATO Emerging Security Challenges Division.

Sub-Series

A.	Chemistry and Biology	Springer
B.	Physics and Biophysics	Springer
C.	Environmental Security	Springer
D.	Information and Communication Security	IOS Press
E.	Human and Societal Dynamics	IOS Press

- http://www.nato.int/science
- http://www.springer.com
- http://www.iospress.nl

Series B: Physics and Biophysics

Medical Response Strategy in Case of Radiation Emergency Caused by the War in Ukraine

edited by

Florin-Catalin Cirstoiu
Bucharest University Emergency Hospital
Bucharest, Romania

Victor Juc
Moldova State University
Chișinău, Moldova

Corina Silvia Pop
Bucharest University Emergency Hospital
Bucharest, Romania

Petre Min
National Commision for Nuclear Activities Control
Bucharest, Romania

and

Cristian Barna
Bucharest University
Bucharest, Romania

NATO OTAN *This publication is supported by:* The NATO Science for Peace and Security Programme

Springer

Published in Cooperation with NATO Emerging Security Challenges Division

Proceedings of the NATO Advanced Research Workshop on
Regional Strategy for Medical Response as part of the Disaster
Management in Case of Radiation Emergency caused
by the War in Ukraine
Bucharest, Romania
19–21 September 2023

ISBN 978-94-024-2268-9 (PB)
ISBN 978-94-024-2265-8 (HB)
ISBN 978-94-024-2266-5 (eBook)
https://doi.org/10.1007/978-94-024-2266-5

Published by Springer,
P.O. Box 30, 3311 GX Dordrecht, The Netherlands

www.springer.com

If disposing of this product, please recycle the paper.

Preface

Russia's military invasion of Ukraine is considered the most serious security crisis in Central and Eastern Europe since the end of the Cold War, with major implications for the regional security environment and NATO.

Since the very beginning of Russia's illegal military invasion of Ukraine, the Heads of State and Government of the member countries of the North Atlantic Alliance condemned the invasion and demanded that Russia should stop and withdraw its forces from Ukraine, being deeply concerned that the violence and insecurity in the region caused by Russia are determining a critical humanitarian situation and material destruction all over Ukraine.

Moreover, after invading Ukraine, Russian army took control of the Zaporizhzhia nuclear power plant, in early March 2022. Since then, Russian military units have guarded the facility and, after Russian President Vladimir Putin annexed up to 18% of Ukraine, including the area where the nuclear plant is located, he signed a decree to formalize Russian control over the plant.

In this respect, we must underline that each of Ukraine's operating nuclear power reactors and the high-level radioactive waste stored near them contain massive amounts of hazardous and long-lived radioactive material. Direct attack on their operation, including interruption of the electrical power and water needed for constant cooling of both reactors and spent fuel ponds, would risk fires and explosions that could cause a radiation accident.

Also, a series of civil institutions—such as hospitals, research institutes or factories that legally used radioactive sources, were bombed and devastated, raising legitimate concerns about environmental contamination, theft or mishandling of the radioactive sources.

In the case of a major accident, radioactive debris would likely spread to countries surrounding Ukraine, and depending on weather patterns, that could well include Russia, as well as to the rest of Europe, possibly even to the Middle East and North Africa.

Given the above mentioned, this book addresses a potential radiation incident caused by the war in Ukraine, from an interdisciplinary approach of medical, nuclear safety and security, nuclear research, geostrategic and population resilience perspectives. This book also addresses the key role played by NATO in identifying new solutions for stabilizing the region, while also focusing on the emergency preparedness in order to ensure that adequate capabilities and capacity are in place which can be mobilized in response to a radiation emergency.

Bucuresti, Romania — Florin-Catalin Cirstoiu
Chişinău, Moldova — Victor Juc
Bucuresti, Romania — Corina Silvia Pop
Bucuresti, Romania — Petre Min
Bucharest, Romania — Cristian Barna

Contents

About the Editors

Florin-Catalin Cirstoiu MD, PhD is Manager of Bucharest Emergency University Hospital, Professor of Orthopedics and Traumatology and Dean of Medicine Faculty at "Carol Davila" University of Medicine and Pharmacy Bucharest. Furthermore, he is the Chief of the Orthopedics and Traumatology Department at Bucharest Emergency University Hospital where he conducts extensive surgical activity centered on bone defects reconstruction using various methods after tumor resection surgery, but also after failed total hip or knee arthroplasty and osteolysis caused by specific or nonspecific osteoarticular septic complications.

Victor Juc is Head of the Legal, Political and Sociological Research Institute, Moldova State University, and Doctor, Professor, Corresponding Member of the Academy of Sciences of Moldova, scientific profile International Relations. Professor Victor Juc is author of 310 scientific publications (monographs, chapters in collections, articles in scientific journals, materials of national and foreign scientific conferences).

Corina Silvia Pop is Professor of Internal Medicine and Gastroenterology at Carol Davila University of Medicine and Pharmacy Bucharest—over twenty years of clinical and academic experience Medical Director of University Emergency Hospital Bucharest Secretary of State at Ministry of Health 2015-2018 Senior adviser of Ministry of Health Senior advisor health projects Ministry of European Funds Between 2015-2018 she coordinated National Health Programmes financed by European Social Funds EU Commission on Communicable and Non-Communicable Diseases screening and prevention.

Petre Min is a nuclear power plants engineer, with an MSc in Radiation Protection and Nuclear Safety and a PhD in NPP Accident Management, with 19 years of working experience. His duties at the regulatory authority included the coordination of Emergency Preparedness and Response activities, implementation of Emergency Preparedness and Response projects, elaboration of regulatory requirements in Emergency Preparedness and Response, Radiation Protection and Nuclear Safety, elaboration of plans and procedures, national representation at international organizations (EC, IAEA).

Cristian Barna is Training Manager at Intelligene4ALL, Visiting Professor at "Babes-Bolyai" University from Cluj-Napoca and Bucharest University, and former Vice-Rector and PhD coordinator of "Mihai Viteazul" National Intelligence Academy from Bucharest. Cristian Barna holds a PhD in Sociology, and his areas of expertise are applied sociology, intelligence and security studies, geopolitics, security organizations, organized crime and terrorism studies.

Part I
The Risk of a Radiation Incident Generated by War in Ukraine. Implications for International Actors and Regional Security

The End of the War: The Worst Case Scenario for Europe

Ștefan Dănilă

The war triggered by Russia by invading Crimea is far from over, the outcome of this war is uncertain, although there are many predictions and assumptions. The scenarios regarding the evolution of the war and the expected ends differ from the current of thought to another, from individual to individual, depending on vision, evaluation model, involvement, and interest.

From the beginning, the US decided to take joint, firm action, in conjunction with those of its allies (mainly the democracies of the world), to keep the war at the level of a limited conflict. The positioning on Ukraine's side, namely the aggressed side, did not imply direct involvement, and this decision implied a special effort, with modest but encouraging results. Overcoming evil by non-lethal means, economic and political measures became the main objective of the US, but realist thinkers saw in this decision a weakness, an inability of the US to impose its role as world leader. After several assessments, since the first 6 months of the conflict, we have identified at least seven scenarios, some similar to others, but differentiated by consequences.

I will expose the determining hypotheses of the production of each scenario, as well as the consequences of each resulting end in order to understand which is the most desirable scenario (the optimistic version), but also the worst scenario for the European Union, implicitly for Romania, but also the worst for Romania, without an implicit Community variant. I consider it very important to know what future we expect and to identify the best actions to achieve a high level of national security under the new conditions.

My conclusion, not necessarily that of all those who follow the unfolding of events, is that Romania is not yet ready for a neighborhood with a democratic Ukraine, partner (future member of the EU and/or NATO), but it must be. A different choice could push Romania into isolation and hostility at the border.

Ș. Dănilă (✉)
Integrated Intelligence Defence and Security Solutions, Bucharest, Romania

F.-C. Cirstoiu et al., *Medical Response Strategy in Case of Radiation Emergency Caused by the War in Ukraine*, NATO Science for Peace and Security Series B: Physics and Biophysics, https://doi.org/10.1007/978-94-024-2266-5_1

It's been almost a year since the announced but hard-to-believe invasion of Ukraine by Russian armed forces. Predicted, warned, perhaps even expected by some, the "Operation to Denazify and Demilitarize Ukraine" began on the morning of February 24, 2022, a date that I wish would not become as important in human history as September 1, 1939.

Ukraine and possible scenarios are talked about a lot, at least in Europe being the most important topic. It is worrying that our lives have been and are influenced by what is happening not too far from Romania, all the more so as we do not have a direct contribution to the outbreak of this conflict and we have felt from the first days the consequences of a war nearby.

Energy crisis, inflation, rearmament, a NATO concept with a significant increase in forces prepared for war, food crisis warnings, arms flows, heated debates in TV studios and on social networks, generals, presidents, prime ministers, or defense ministers, all these were consequences of a military aggression that aimed to "restore the historical right of the Russians".

The consequences of war affect us more and more, in various forms, from filling up the car, to electricity and gas bills, from the liter of oil or sugar to bank rates and giving up holidays, to losing our job, car or house, to accepting (in the beginning it was support!) the thousands of refugees we meet on the streets, in shops, in traffic. We received requests for support and offered support.

Disapproval of the aggression of a large state against a smaller state is quasi-general, but, as with any prolonged effort, fatigue is felt. More and more people want this war to stop, no matter who has to give in. A violent conflict can be prevented by deterrence only when both sides realize that they have more to lose by starting a war, but the situation is different after the onset of violence. It is known that once armed confrontation begins, a return to negotiations requires at least one side to feel that it has more to lose if fighting continues or the war continues until actual defeat.

The exhortation to the Ukrainians to give in seems to obscure the idea that Ukraine will not be able to defeat Great Russia, that David will not be able to defeat Goliath. Increasingly, the "right of the fittest" is being invoked even by renowned theorists such as renowned professor John Mearsheimer or former Secretary of State Henry Kissinger, who already blame the US and NATO for fueling Ukraine's impossible dream of NATO membership. And this trend has more and more followers, amid the feeling of the consequences mentioned. Beyond rights and principles, one's own needs are more important, the sacrifice of well-being cannot be supported with words, flags and videos, even if the latter reveal cruelty and destruction. Victims become guilty, and any additional demand touches sensitivities among more and more citizens, with priority given to those in neighbouring states.

The warnings sent by intelligence services, especially US military intelligence during 2021, were very accurate and generated serious analysis before the war began, and some were made public, others were not. We now know about the military recommendations made by the Chairman of the Joint Chiefs of Staff of the US Armed Forces, General Mark Milley, as quoted at the beginning of this text, following information received in October 2021.

Since the beginning of the war, "red lines" have been identified, meaning actions that could lead to escalation, both from the Russians and from the democratic world, of the states that stood by Ukraine, identified by the Russian leadership as the US, NATO, and the EU. The US imposed sanctions on the Russian Federation and some Russian citizens, measures also assumed by the European Union, shortly after. This was followed by support with equipment, ammunition, and recently even heavy weapons, modern tanks.

Analysis of the situation of the front, up to date, is a topic in the program of all news stations, there are more and more experts more or less knowledgeable who know in detail every tactical move. Unlike other states, which understood the importance of strategic communication from the beginning of the war, the Romanian domain responsible, namely the Ministry of National Defense, remains in the shadows, after some at least funny outings. The vanity of appearing on screen and the desire for notoriety are great, and the place left vacant by those who have the responsibility of communication to prevent intoxication with fake news or misinformation must be occupied by someone!

I will avoid getting into the game of the "service expert" who makes the analysis of the combat forces, of the available or presumed resources, I will limit myself to making a simple description of possible scenarios, without claiming to have identified them all, as well as the possible consequences for Romania, so that it is clearer on which side we stand. I will also avoid quotations or juxtapositions that generate syntax analysis, and their mere order leads to the labeling as "pragmatic", that is, eager to end the war at any cost, or "idealist", supporter of defending democratic values at any cost, or pro-American, much less likely, "Putinist". This approach is meant to help guide individual actions and attitudes, but also at state level towards the most appropriate direction for Romania, for our interests.

The first possible and, I believe, necessary scenario to be considered would be that of cession by Ukraine. The probability of this scenario may increase against the background of destruction of critical infrastructure, loss of life (not only at the front, but also among the civilian population, sometimes quite far from the front), lack of means of combat. In the first phase it could mean ceding positions and surrendering frontline forces, gradually and progressively until surrender, a situation with quite low probability at this point, it could be the result of a strong mass movement determined by ceding the will to fight among the population (and this, still unlikely), but also by liquidating the Ukrainian political leadership through a lightning action of Russian special forces, followed by a spontaneous popular anti-war movement.

The first hypothesis would be possible against the background of physical exhaustion of the military, lack of ammunition and the start of an offensive by Russian forces. The strong motivation of the military defending their country, supported by the effective support with ammunition, equipment and weapons from the states supporting Ukraine make this hypothesis only a theoretical one.

The second hypothesis of the popular movement against war would have a major impact on the troops on the front, could overturn the situation and could mean the fall of the current political leadership and the establishment of a leadership not necessarily pro-Russian from the first moment, rather anti-war, eager to conclude peace.

The consequences could mean an indefinite truce and accelerated peace negotiations. Russia will not cede the territories already seized, on the contrary, it will demand the territory of the four regions as they were drawn by their so-called representatives in the Moscow show on September 30, 2022. Ukraine should withdraw its applications to join NATO and the EU, as well as amend its constitution accordingly. Ukraine will have to annul its education and minority laws. It could also be possible for Russia to request the placement of Russian bases on Ukrainian territory in response to the presence of NATO troops near Ukraine's border. Ukrainian public opinion could turn hostile towards NATO and the EU as a result of frustrations generated by the lack of direct involvement of NATO and the EU. Very likely, the political leadership will change in Moldova with pro-Russian forces.

The Russia-NATO relationship will be that of the Cold War, with sanctions and tensions remaining. The Russia-EU relationship will be equally tense, the EU will continue to look for solutions so as not to depend on energy and mineral resources from the Russian Federation. China will be the main beneficiary of peace, engaging in reconstruction and securing the market for Russian and Ukrainian products.

Romania would return to the status of NATO and EU border state under the conditions of tense relations with Russia. Relations with Ukraine could become extremely tense as a result of the Ukrainian public opinion diverting attention from its previous enemy, Russia, to the west, and Romania is among the most vulnerable.

The third hypothesis, that of removing by force the Ukrainian political leadership and imposing a pro-Russian regime would have much the same consequences, maintaining a potential anti-Russian attitude among the Ukrainian population. Imposing Moscow's conditions would be very fast, but the probability of starting Ukrainian popular uprisings and maintaining the pro-Western attitude would be high.

The second possible scenario, without establishing an order according to the level of probability, would be the abandonment of the fight of the Russian armed forces, exhausted, without real motivation, influenced by the Ukrainian offensive, by the will of the Ukrainians to fight for their own country. It could happen gradually, becoming contagious amid the existence of the Russians' own social network. Russian commanders would lose control of the troops, and they would surrender, under the conditions of humane treatment, over the conditions of prisoners of war, and even with transport to the desired area/country.

The consequences could be the rapid change of the front line, the reconquest of vast territories, except Crimea and, perhaps, the Donbass area. The Russian Federation would trigger general mobilization and could be open for negotiations and a temporary truce. Social unrest could arise in Russia, leading to the establishment of a more radical, nationalist government. Strikes on the territory of the Russian Federation and Belarus would continue, the conflict would remain open. Russia will not declare itself defeated. States supporting Ukraine would have time to strengthen Ukraine's defense capability. As well as to support the reconstruction of Ukraine. Sanctions against Russia will remain, with the same economic implications. China will benefit from Russia's resources, but it may have a reserved attitude towards the conflict.

The Republic of Moldova could maintain its European course, strengthen itself with the support of the Union, and the Russian Tactical Task Force could be put under pressure with the support of Ukrainian forces, which would isolate any communication route with Russia, could be evacuated or dismantled. Romania could contribute to strengthening defence capacity, as well as to the reconstruction of Ukraine, could provide assistance to accelerate the process of its accession to the European Union.

Another possible scenario is that of "freezing the front line", namely the involvement of the US and the EU in peace negotiations with the Russian Federation, with Ukraine in the background. This scenario would be possible under great pressure from its own citizens, amid the depletion of Ukraine's resources to cope with the Russian offensive. It could involve obtaining a ceasefire agreement on a certain alignment. This alignment could be a demarcation line between the European Union and Russia. That is, Russia will maintain the conquered territories and there is a possibility that Russia will no longer recognize Ukraine as a state, as a subject of international law. The remaining territory of Ukraine could become part of the EU, recognised by almost all UN states except Russia and possibly its allies, who could discuss any Ukraine-related issue only through EU representatives.

The consequences would be disastrous for Ukraine but could mean restored economic relations between Russia and the EU, even if Cold War hostility were present. For the U.S. it could mean a withdrawal from Europe to solve domestic and Asia-Pacific problems, but that probability is extremely low at this point. For the Republic of Moldova and Romania, the consequences would be similar to the previous scenario.

Another scenario would be the cessation of military support for Ukraine, possibly under pressure from domestic public opinion, without involvement in the peace process or not accepted in peace negotiations by the Russian Federation. From this scenario one can easily enter into the hypotheses of the first scenario, which would lead to Russia's victory, with all other consequences.

The worst-case scenario, but one that cannot be neglected, is that of direct US involvement in the conflict, by creating a coalition of will to effectively support Ukraine in the fight with forces. There are minimum prerequisites for such a coalition through the group of states participating in Ukraine's arms support, the Rammstein Group, as well as from the moral obligation of the signatory states to the Budapest Memorandum of 1994, which guaranteed Ukraine's territorial integrity in exchange for ceding nuclear weapons from its territory, Russian Federation. These states are the US and Great Britain, and this scenario is very likely given the use of nuclear means against Ukraine, as this situation is clearly prohibited in the Memorandum.

The consequences could be disastrous. In a first stage, Russian forces could be relatively easily defeated by the alliance through a joint and combined operation, Crimea could also be liberated within weeks, and Ukraine's pre-March 2014 borders would be restored. Russia's reaction could be difficult to control, as it would invoke the situation of endangering the existence of the state and could use nuclear weapons against participating states. The greatest danger is the launch of a second missile or nuclear bomb, that is, the US response or Russia's insistence. Mutual destruction would be extremely difficult to stop. China is likely to position itself openly against a coalition against Russia, which could mean confirmation of the start of 3rd World War. Romania is in the beating of short- and medium-range ballistic missiles, and the Deveselu missile system does not protect us. Missile defense is almost non-existent at the moment, which would make any important objective in Romania an easy target for Russia.

There is also a scenario in which the parties would accept a ceasefire and a circumstantial peace, with a negotiated demarcation line, with tensions and conflict keeping at a low level, towards frozen, but with the détente of economic relations between the EU and Russia, with ensuring the transfer of energy, oil and gas from Russia to Ukraine through guarantees or contracts of the European Union. For this scenario, both sides would have to realize that any hour of war means greater destruction and loss of life, and for Ukraine it would mean accepting the loss of particularly important territories.

In international relations, in the reality of the world we will live in, such a scenario would mean an accelerated arms race and very likely the reopening of many conflicts worldwide, based on the principle of imposition by force. It would mark the bankruptcy of US hegemony, its inability to rule the world on the principles set out in the UN Charter. The UN would become obsolete, and the EU could experience destructive turmoil. There are many analyses, referring to already existing situations, from post-World War II peace treaties to those in Korea, but it is hard to make predictions for the end of a war.

The seventh scenario is that of "NATO involvement in the defense of a member state", which implies a possible attack on a NATO member state by the Russian Federation. Favorable factors could be mutually hostile messages, the fact that the main reason for Ukraine's aggression is to stop joining NATO, NATO states neighboring Ukraine support and ensure the transport of war aid, Russia wants to stop this aid.

It could be achieved by: hitting a communication route, port or airfield of a NATO state (Poland, Estonia, Lithuania or Romania) by Russia, or hitting a plane in Ukrainian airspace or a NATO state ship in international waters. NATO would respond jointly by conducting an air and naval operation aimed at achieving an NFZ (Non-Flight Zone) and/or a safe maritime zone.

The consequences could be hard to manage. In a first stage, NATO would intervene directly against Russian forces—air strikes on targets in occupied territory. Russia, as it loses ground, will likely threaten nuclear weapons and even strike a U.S. naval group (Sixth Fleet?) or some other isolated target with a nuclear-tipped missile.

A similar US retaliation is possible, with the destruction of important Russian targets (nuclear submarines, aircraft carriers). The first retaliation may not be nuclear. If negotiations are not launched amid a temporary truce, mutual nuclear strikes will be triggered and states close to Russia will be the most vulnerable, the most likely targets will be US military facilities in Romania, Poland and the Baltic States.

Turkey could withdraw from NATO. The outcome can be unpredictable, from accepting peace with the liberation of the entire territory of Ukraine, under the conditions of lifting sanctions, returning to the NATO-Russia Agreement, giving up Ukraine's NATO membership and Russia's acceptance of Ukraine's EU membership, to the mutual destruction of Russia and most NATO states through the use of nuclear weapons on a large scale. China could evade conflict or force a military invasion of Taiwan, with the consequence of going to war with the US. For now, the only certainty is that war causes death, destruction and collateral damage to the entire world.

Ukraine will face big problems in either scenario amid demographic problems. The war dead and massive westward migration have reduced the population by almost a third, and the prospect of a significant number returning to the country is only raised for a scenario in which at least part of Ukraine remains under the control of the European Union. In the other scenarios, Ukrainian refugees will remain in the European states where they will settle, and this could be a growing problem for the European Union, which is also facing the migration flow from the south. There may be a clash of immigration flows, as well as the rise of radical, right-wing, anti-migration nationalist movements. The entry of Romania and Bulgaria into Schengen seems to be an increasingly distant dream, regardless of the overly optimistic messages of Romanian politicians. So, Pandora's box has been opened, the hope that wisdom will find solutions to return to a state of peace is diminishing with each day of confrontation.

European Security in the Context of the Russian Federation's Extended War Against Ukraine

Victor Juc

Abstract The article delves into European security dynamics against the backdrop of the armed conflict initiated by the Russian Federation against Ukraine. It specifically examines the interplay between the European Union and the North Atlantic Treaty Organization, the evolutions in the eastern Euro-Atlantic region, and the resurgence of discussions surrounding the potential utilization of nuclear armaments by certain states.

The overarching conclusion suggests that the European security framework is experiencing transformative shifts, raising the possibility of a potential emergence of a new "iron curtain."

1 Introduction

The examination of European security within the contextual framework of assessing the prolonged conflict initiated by the Russian Federation against Ukraine constitutes a subject matter of significant scientific and applied scientific inquiry. This protracted conflict spans two distinct phases: initially, a passive phase commencing in the latter half of 2013, characterized by the utilization of hybrid tactics to coerce Ukraine into refraining from signing the Association Agreement with the European Union. Subsequently, in the spring of 2014, the conflict transitioned into an active phase marked by the annexation of the Crimean Peninsula and the backing of territorial and political separatism in two regions.

The onset of the second period on February 24, 2022, witnesses the escalation of the active phase of the armed conflict by the Russian Federation, with objectives encompassing the erosion of Ukrainian national identity and sovereign rights, the imposition of a puppet government regime in lieu of the democratically elected one, territorial disintegration of the Ukrainian state, and the negation of its political independence. The Russian Federation perceives Ukraine as an integral component

V. Juc (✉)
Moldova State University, Chişinău, Moldova

F.-C. Cirstoiu et al., *Medical Response Strategy in Case of Radiation Emergency Caused by the War in Ukraine*, NATO Science for Peace and Security Series B: Physics and Biophysics, https://doi.org/10.1007/978-94-024-2266-5_2

of the Russian world and a crucial element in its pursuit of great power status, echoing assertions articulated by Brzezinski [1] and Friedman [2].

The relevance of the research topic is underscored by the imperative to scrutinize the evolving dynamics within the European security landscape, amidst anticipated shifts in the prevailing world order paradigm and the structure of the post-Cold War international system. The significance of this research endeavor is driven by the necessity to comprehensively understand and elucidate the milieu, conditions, contextual factors, and trajectories pertaining to the evolution of European security. Such understanding is indispensable for identifying pragmatic solutions aimed at mitigating risks and fostering the development of post-conflict scenarios. Conceptually, the trajectory of international relations is marked by cyclical shifts between phases of cooperation and of conflict.

These alternations are precipitated either by transformations in the prevailing world order model, as exemplified by the conclusion of the Cold War in the late 1980s and early 1990s, or by shifts in the international context, exerting profound influences on the nature of global interactions. We posit that post-Cold War international relations have traversed two distinct phases: the initial period spanning from the late twentieth century to the middle of the second decade of the twenty-first century, and the ongoing second period since then. The first phase is distinguished by a prevailing ethos of cooperation within the uni-multipolar model of world order in geopolitical terms, alongside a polycentric structure characterizing geoeconomic relations. In contrast, the second period is characterized by heightened conflict dynamics within the uni-multipolar model of world order, with emerging tendencies towards multipolarity or bipolarity and a resurgence of geopolitics overshadowing geoeconomics. This phase culminates notably in the aggressive actions of the Russian Federation against Ukraine on February 24, 2022.

The security analysis employs the "danger" category, which encompasses two distinct types: risk, representing potential danger, and threat, denoting real danger. The working hypothesis posits the following: European security presently confronts significant risks, and there is a lack of identifiable conditions indicating an imminent transition from risk to threat. The scientific research methodology hinges on the actor-structure relationship, whereby both influential actors and the overarching structure collectively shape and adapt a particular type of international system. Among major powers, including superpowers and great powers, the configuration of the international system is predominantly influenced by established actors leveraging their available resources.

Meanwhile, structural shifts serve to dictate alterations in the conduct and interactions among these actors. The endeavours of international actors unfold within specific contexts, which may either facilitate or hinder the development of conflict or cooperation relations [3]. In delineating the array of methods employed within the scientific research process, it is pertinent to highlight the utilization of various methodologies, including the institutional method, systemic approach, structural-functional approach, behaviourist method, as well as techniques such as analysis and synthesis, content analysis, event analysis, and the forecasting and scenario construction method.

2 European Security Within the Context of the Partnership Between NATO and the EU

The unprovoked aggression perpetrated by the Russian Federation against Ukraine on February 24, 2022 has reignited discussions surrounding the imperative of bolstering Europe's military and political security. Compounding this challenge is the potential shift in strategic foreign and defence policy options by the United States of America (USA). It is noteworthy that most European states have historically afforded less priority to security and defence policies, relying instead on active participation from the United States and the North Atlantic Treaty Organization (NATO). Additionally, the confidence among European political elites that armed conflicts are incompatible with the democratic geopolitical landscape of Europe, influenced by the prevailing notion of triumphalism elucidated by F. Fukuyama [4], has further diminished attention towards military matters.

It is important to clarify that while wars have not been entirely absent from the European space, they have been predominantly perceived as localized and peripheral, confined to the former Soviet and former Yugoslav regions without expanding to a broader European regional scale. Western European states have progressively reduced funding allocated to national security and defence sectors, encompassing areas such as military industry, armed forces, reservist capabilities, equipment procurement, and research initiatives. During the initial period of the armed conflict in Ukraine (2014–2022), there was only a marginal influence on security and defence policies. President Donald Trump's stance in 2017 advocating for increased budgetary contributions within NATO was widely perceived by Europeans as unjustified.

It is crucial to underscore that while the revisionist stance of the Russian Federation concerning the post-Cold War model of world order and international security architecture has been acknowledged, the European arena in 2022 does not manifest a conducive environment for the outbreak of a large-scale war in Europe. Consequently, the escalation of the armed conflict in Ukraine could have been averted. The incursion by the Russian Federation necessitates European nations to not only reinforce existing institutions but also establish new platforms, given the vulnerabilities in terms of sustainable preparedness for a prolonged conflict. In August 2022, President Vladimir Putin evoked the Northern War, which endured for 21 years, with 7 years characterized as "frozen." This historical reference underscores the predisposition of the Russian Federation to engage in protracted armed conflicts, ostensibly invoking the fight against NATO as justification.

Undoubtedly, throughout the Cold War era, the security of the European populace within the Western cultural-civilizational geopolitical sphere was effectively safeguarded by NATO. Following the conclusion of the bipolar conflict, the number of nations expressing the desire to integrate into the Euro-Atlantic realm under the principles of democracy, rule of law, freedom, economic collaboration, and collective defence substantially increased. NATO facilitated the establishment of a unified defence framework by enhancing both individual and collective capabilities to deter and counter armed aggression.

It is worth emphasizing that the cohesion within the Euro-Atlantic region is upheld through the commitment of each member state to refrain from assuming any international obligation that contradicts the principles outlined in the Washington Treaty. Additionally, member states pledge to engage in consultations whenever a member perceives a threat to its territorial integrity, independence, political sovereignty, or national security.

Furthermore, immediate support, whether individual or collective, is extended to member states in the event of an armed attack. Through consultation and mutual support, the security of the Euro-Atlantic space is bolstered, while deterrence strategies and interoperability, particularly in the utilization of multinational forces, further strengthen the sense of solidarity. It is these shared values and collaborative measures that underpin the efficacy of the North Atlantic Treaty.

Referring to the European Union (EU), it is pertinent to highlight that the Common Security and Defence Policy, established by the Treaty of Lisbon as an integral component of the Common Foreign and Security Policy, serves as a policy framework facilitating cooperation and coordination among member states in defence matters. This framework not only contributes to crisis management but also enables the EU to undertake civil and military missions and operations abroad [5]. It is noteworthy that the EU has adopted and is actively implementing various strategies and initiatives aimed at enhancing the effectiveness of its foreign policy, security, and defence efforts. Key priorities include bolstering the resilience and capacities of partner nations, adopting a comprehensive approach to conflicts and crises, pursuing strategic autonomy, and safeguarding the European space and the citizens of both member states and third countries alike.

A recently adopted initiative is the "Strategic Compass," a document in which member states outline guidelines for the EU's common security and defence policy. The Strategic Compass aims to foster a shared understanding of the imminent dangers and challenges confronting Europe in the short and medium term. The common security and defence policy is organized around four key pillars: security, partnerships, investment, and action [6]. It is essential to emphasize that a robust and capable European Union in terms of security and defence is complementary to NATO, which serves as the cornerstone of collective defence for the Euro-Atlantic states.

The conflict initiated by the Russian Federation in Ukraine has prompted discussions on various approaches toward effectively defending Europe against the potential expansion of Russian aggression. An equally significant operational concern pertains to the strategies, tactics, and resources required to achieve this existential objective. The need for enhanced cooperation and delineation of responsibilities and actions between NATO and the EU has resurfaced, notwithstanding the fact that 22 European states are members of both organizations. Clarifying the institutional delineation of roles to mitigate overlap and enhance efficiency proves crucial for both multilateral cooperation bodies. Since the adoption of the EU-NATO Declaration on the Common European Security and Defence Policy in 2002, partnership and cooperation between the two institutions in crisis management have steadily progressed. Through the Berlin Plus package, the parties have agreed that Euro-Atlantic forces and assets can be deployed in operations under the auspices of the EU [7].

It is noteworthy that the Berlin Plus agreements explicitly facilitate joint exercises in crisis resolution, encompassing the transfer of responsibility from NATO to the EU and the exchange of information based on mutually agreed security norms. Through the Declaration of June 28, 2016, the European Council within the EU emphasized the necessity for deeper collaboration between the EU and NATO, grounded in the shared objectives and values of both organizations, while also respecting their decision-making autonomy and respective governance procedures. Notably, the EU-NATO Joint Declarations of 2016 and 2018 delineated strategic areas of cooperation, including hybrid threats, cyber security, industry and research, coordinated exercises, defence capabilities, military mobility and combating terrorism [8].

The invasion carried out by the Russian Federation in Ukraine was met with vehement condemnation from both NATO and the EU, leading to intensified cooperation efforts between the two entities. In the NATO-EU Joint Declaration of January 10, 2023, both parties reaffirmed their unwavering support for Ukraine and articulated a shared agenda of activities aimed at addressing common security threats.

Areas of cooperation outlined in the declaration encompassed geostrategic convergence, critical infrastructure resilience and protection, emerging technologies, climate change, and countering foreign manipulation of information [9].

Henceforth, these Declarations underscore the deepening of the strategic partnership between NATO and the EU, rooted in shared values and a collective resolve to confront challenges to peace and security. Safeguarding European and Euro-Atlantic security and stability through collective defence necessitates significantly closer and more inclusive cooperation between NATO and the EU, utilizing the tools at their disposal. While both multilateral cooperation organizations maintain unequivocal strategic decision-making autonomy, technical agreements and arrangements approved by each party may be mutually applicable. As cooperation between NATO and the EU continues to deepen and expand, it becomes imperative for both parties to complement each other in terms of delineating and sharing spheres of action.

Simultaneously, European states are increasingly inclined towards bilateral initiatives to address security and defence concerns. For instance, France and Germany are actively developing policies aimed at enhancing European defence capabilities in response to emerging realities and trends in the European security landscape. However, divergent positions exist regarding the strategic and tactical preferences of the two nations: France advocates for an autonomous European defence posture, reminiscent of neo-Gaullism, while Germany advocates for the consolidation of transatlantic relations. Another pivotal element of Europe's collective defence policy is the "Weimar triangle" involving France, Germany, and Poland. This framework serves as both a mechanism for ensuring defence cooperation and a platform for identifying solutions to conflict resolution.

3 The Reinforcement of NATO's Eastern Flank as a Critical Imperative Considering the Proliferation of Threats to European Security Since 2014

The North Atlantic Treaty Organization (NATO) emerged from the collective aspirations and security imperatives of its founding European and North American states, committed to providing mutual support through the principle of collective defence. Following the collapse of the "iron curtain," the desire for Euro-Atlantic integration among some former communist states surged, driven by aspirations to bolster their national security and contribute to the broader strengthening of Euro-Atlantic security. The London Declaration of July 6, 1990, publicly announced NATO's transformation programs, which included initiatives aimed at collaborating with former adversaries to enhance European security. These efforts materialized through the promotion of the "open door" policy and the initiation of pre-accession and accession programs, such as the Partnership for Peace. It is noteworthy that the Republic of Moldova has been a participant in this program since March 16, 1994, signifying its commitment to fostering cooperative security measures within the Euro-Atlantic framework.

The eastern flank comprises Romania, Bulgaria, Hungary, Slovakia, Poland, Estonia, Latvia, and Lithuania. These nations are situated near the Russian Federation and confront security and national defence challenges, which have been significantly heightened following the commencement of the unjustifiable act of aggression against Ukraine. The proliferation of threats in the region necessitates a continuous focus on deterrence and countermeasures. In strategic documents developed by NATO in 2022 and 2023, the Russian Federation is no longer regarded as a partner but rather as an adversary, prompting the adoption of strategies and tactics tailored to address the evolving realities. At the North Atlantic Alliance summit in Madrid on June 29, 2022, three primary objectives related to ensuring collective defence were identified: deterrence and defence; crisis prevention and proactive crisis management; and cooperative security [10].

The accession of central and southeastern European states to Euro-Atlantic institutions has naturally afforded them political-military protection and deterrence against threats from the Russian Federation. However, since the onset of the Russian Federation's military aggression against Ukraine in 2014, involving territorial disintegration and support for secessionism, eight battalion-level battle groups have been deployed in NATO's eastern flank.

These battlegroups, deployed in the Baltic States and Poland in 2017, with the remainder stationed in 2022, are tailored to the sub-regional military situation, with force size and military capacity varying based on geographical factors and the specific risks and threats faced by each Eastern Flank country. At the NATO summit in Madrid, it was decided that if there arises an operational need to deter potential adversaries, the battlegroup forces would be augmented and fortified with necessary military equipment. These forces would be further supported by significant reinforcements, readily available to intervene with equipment, weapon systems, and additional supplies, including fuel. To facilitate such actions, adaptations to command-and-control structures are deemed necessary, alongside the establishment of structures at the division level. Regarding additional military forces, while not all would be permanently deployed in the sub-region, they would be assigned to specific territories within the eastern flank area. These forces would undergo rotational deployments, ensuring the development of on-site capabilities and interoperability with host countries' forces through an efficient rotational mechanism.

Currently, the eastern flank stands out as the most exposed and vulnerable segment of NATO, facing significant threats and risks emanating from the Russian Federation. Within the eastern flank, three central points of strategic importance are under considerable pressure, necessitating NATO's accurate assessment of the levels of danger associated with each and the implementation of coherent policies and robust deterrence measures. To the north lies the Baltic Sea region, while the central area encompasses the Suwalki Corridor and Poland, both of which are vulnerable to potential aggression. In the south, the Black Sea region presents additional challenges.

Strengthening NATO's military presence in the southern area, particularly in the extended Black Sea basin, is crucial for regional security and deterrence with global implications. This includes initiatives such as the modernization of the Mihail Kogălniceanu base in Romania. It's worth noting that Romania, positioned on the southern segment, and Poland, located on the northern segment of the eastern flank, are emerging as pivotal actors in bolstering regional security and contributing to NATO's collective defence efforts.

The extended geopolitical area of the Black Sea basin holds significant strategic importance, and following the events of 2014, NATO embarked on the construction of a network comprising four multinational tactical battlegroups as part of the Enhanced Forward Presence (EFP) initiative. Post-2022, the Alliance intensified its efforts and established an additional four battlegroups in Romania, Bulgaria, Hungary, and Slovakia, further bolstering its missions and presence in the region.

The geopolitical doctrine of the Russian Federation firmly rejects the prospect of NATO encroaching upon its European and South Caucasian borders. The incorporation of the Baltic States into the European geopolitical and cultural-civilizational sphere was perceived by the Russian Federation as a blow to its status as a great power and a reduction of its influence in maintaining former Soviet republics within its geopolitical orbit. From the perspective of the central authorities of the Russian Federation, the exclusion of Russia's geopolitical influence from countries like Ukraine and Georgia would result in NATO's encirclement of its borders, the elimination of buffer zones, and an intolerable threat to its national security and that of its allies.

In its efforts to bolster its presence in Central and Southeastern Europe, the Russian Federation employs various methods, primarily of the hybrid nature, though the use of conventional or even tactical nuclear weapons cannot be entirely discounted. Skilfully leveraging the ruling elites of certain Euro-Atlantic states, including those on the Eastern flank, the Russian Federation seeks to influence the decision-making process and secure outcomes favourable to its interests.

Despite potential shifts in the policy preferences of some allied states regarding aspects of the conflict in Ukraine, NATO remains the most effective international institution for ensuring security and defence. While NATO does not directly engage in the armed conflict in Ukraine with military forces, it is tasked with fortifying the eastern flank, including through the deployment of military assets in countries situated near the Russian Federation. This ensures the security needs of the states and peoples in this geopolitical area are adequately addressed.

We hold the belief that the conflict instigated by the Russian Federation against Ukraine poses a significant risk of protracted duration, particularly given President V. Putin's expressed aim to dismantle Ukraine as an independent, sovereign, and territorially intact state, effectively eradicating the Ukrainian people's identity and right to existence. It is worth noting that the Russian Federation possesses ample resources to prolong its intervention in Ukraine and is preparing for military actions of a larger scale, potentially involving a confrontation with NATO. This is evidenced by the allocation of approximately 100 billion Euros towards military expenses in 2024 [11].

Hence, the eastern flank emerges as the most vulnerable segment within the Euro-Atlantic space, rendering assistance to Ukraine an existential necessity, given its potential impact on certain NATO member states. However, despite Ukraine's appeal at the Vilnius summit on July 11–12, 2023, to be invited to join NATO, such a request could not be granted. Similarly, the request for security guarantees cannot be fulfilled, as NATO states cannot offer security guarantees to non-NATO states. In the event of a military attack on the Republic of Moldova by a third country, uncertainties prevail regarding whether Romania could intervene with military forces, notwithstanding the significant number of Romanian citizens residing in Moldova.

It's noteworthy that NATO respects the self-proclaimed constitutional status of permanent neutrality, which doesn't hinder maintaining a certain level of partnership and cooperation between the parties. A new milestone in bilateral relations is anticipated through the adapted Individual Partnership Action Plan, set to come into effect on January 1, 2025.

Similarly, it's important to note that Great Britain, under the Security Agreement of January 12, 2024, commits to providing Ukraine with defence assistance packages. This agreement, categorized as a "security agreement" rather than a "security guarantees agreement," entails the provision of additional weaponry across various types, constituting "more assistance." The validity of such agreements in the realm of security and defence will persist until Ukraine's accession to the NATO component. In a comparable fashion, Ukraine has also entered into similar agreements with France and Germany, albeit likely of a smaller scope and extent.

4 The Impact of Nuclear Weapons on European Security

The Russian Federation's ongoing aggression against Ukraine has been accompanied by statements from Russian officials, including both former and current heads of state, regarding the potential use of atomic weapons, both at the strategic and tactical levels. These statements have elicited varied interpretations and understandings, but they have fundamentally reshaped the geopolitical landscape surrounding issues of rearmament and modernization, global control, escalation, and deterrence.

This geopolitical context has fostered conditions conducive to the modernization and expansion of atomic arsenals, often at the expense of disarmament and de-escalation efforts. While threats of atomic weapon applicability are frequent in propagandistic political rhetoric, deterrence remains a present and viable concept.

Consequently, the joint utilization of atomic weapons within NATO, alongside the collective implementation of deterrence strategies, emerges as a crucial means of ensuring collective defence. It's important to clarify that the terms "atomic weapons" and "nuclear weapons" are not interchangeable and carry distinct connotations. While "nuclear weapons" encompasses a broader range of armaments, including atomic weapons, the latter specifically refers to a subset of nuclear weaponry. Additionally, it's worth noting that "nuclear weapons" also encompasses other types of weapons such as chemical, radiological, and biological weapons.

Therefore, while "atomic weapons" specifically denote a subset of nuclear weaponry, "nuclear weapons" encompass a wider array of armaments, including atomic weapons, alongside other types such as chemical, radiological, and biological weapons. Certainly, during the Cold War era, the principle of "mutually assured destruction" effectively deterred the two superpowers from engaging in direct armed conflict, given the substantial risk it posed to the survival of humanity. Similarly, it's plausible to assume that even if Ukraine had retained control over the nuclear weapons remaining on its territory from the Soviet era, the behavior and rhetoric of both Ukraine and the Russian Federation may have been different in the context of the 2014 crisis and the armed conflict of 2022.

It's important to note that during the era of bipolarity, tactics were indeed developed within the Soviet analytical framework to utilize atomic weapons as complementary support in conventional warfare, with the aim of breaching the opponent's defence lines. However, the concept of "mutually assured destruction" effectively ensured the security of states within the "nuclear club" and their allies. In the present context, emerging risks have led some nonnuclear countries to consider acquiring nuclear weapons or exploring joint-use arrangements, akin to those within NATO. For instance, in South Korea, there is a growing inclination towards pro-nuclear state options due to the perceived threat from North Korea. The North Korean regime, possessing nuclear weapons and actively advancing its nuclear program, has explicitly stated its willingness to employ nuclear weapons at the outset of a conventional war.

Conversely, Euro-Atlantic states have consistently affirmed their stance of not being the first to use atomic weapons, emphasizing principles of fairness and responsibility for humanity. This underscores the importance of maintaining a responsible and restrained approach to nuclear deterrence and conflict escalation. Indeed, South Korea, much like Ukraine, opted against pursuing its own nuclear program, instead entrusting the assurance of its security to the United States. However, the aggression of the Russian Federation against Ukraine underscored the vulnerability of a non-nuclear country in relation to a nuclear-armed aggressor, shedding light on the priorities of nuclear states in both strategic and possibly tactical realms. In response, South Korea places significant reliance on deterrence and security in the event of possessing nuclear weapons.

Moreover, the ongoing conflict between India and Pakistan, both nuclear states, vividly illustrates the validity of these assumptions. The conflict remains in a state of flux for short periods without one party gaining clear superiority over the other. Given these considerations, it is understandable that South Korea may seek to reconsider its stance on nuclear proliferation, arguing for the primacy of national interest in shaping its national security and defence policies.

Indeed, the Iranian nuclear program poses a significant threat with potentially enormous destabilizing impacts on international security. The theocratic authoritarian regime in Iran has a track record of promoting armed conflicts through proxies and providing support to organizations declared and recognized as terrorists. The complexity of the situation surrounding the Iranian nuclear program lies in the possibility that nuclear weapons or the capability to manufacture such weaponry could be transferred to terrorist groups. This scenario presents a dual threat, as nuclear weapons could be used not only for deterrence but also for purposes of blackmail and coercion, both at tactical and strategic levels. This underscores the critical importance of international efforts to address and mitigate the risks posed by the Iranian nuclear program, through diplomatic means, sanctions, and other measures aimed at preventing the proliferation of nuclear weapons and technologies to nonstate actors.

While uncertainties surround Iran's nuclear program, the return of Belarus to the list of states potentially possessing atomic weapons is a tangible reality. It's worth noting that until the second half of the 1990s, Belarus housed nuclear warheads of Soviet origin, which were subsequently evacuated. However, the transfer of tactical nuclear weapons by the Russian Federation to Belarus has altered the dynamics of the situation, impacting the environment and state of European security. This development has also bolstered the position of the current Belarusian president. In response, countries like Poland and the Baltic states are likely to undertake deterrent actions to address the heightened security concerns posed by the presence of nuclear weapons in Belarus.

The deterrence crisis is exacerbated by the decentralization of the monopoly on the use of nuclear weapons and the associated uncertainties. Rhetoric surrounding the use of atomic weapons have expanded beyond defensive purposes to include offensive roles as operational tactical tools in conventional warfare. Nuclear escalation now encompasses both strategic and tactical weaponry, with some nuclear states reprogramming for scenarios involving "limited nuclear war." This escalation occurs against a backdrop of reduced efficiency or ineffectiveness of certain international organizations.

The deterrence crisis is further complicated by a lack of trust and progress in developing mechanisms for risk reduction. In the 1960s and 1970s, treaties on non-proliferation and reduction of atomic weapons were signed amidst dialogue between superpowers and strengthened confidence-building measures during periods of détente. However, in the current climate, dialogue is often either impossible or unproductive, and risk reduction measures fall short of what is necessary. The stability of the world order is dependent on both bilateral and multilateral cooperation, with multilateral institutions playing a crucial role in maintaining stability and addressing the challenges posed by nuclear proliferation and deterrence.

We have identified that statements and actions concerning the potential use of nuclear weapons can be classified into three distinct periods:

1. second half of the 1940s: under the conditions of the US monopoly on atomic weapons, it was used as a component of conventional warfare and for intimidation;
2. from the late 1940s to 2022: under the conditions of the increase in the number of states possessing nuclear weapons, it was used for the purpose of deterring, with predilection in the context of the cold war, of the bipolar architecture of international security;
3. from 2022: nuclear weapons are declared to be a tool of intimidation and pressure, while maintaining deterrence capabilities, even if they are in a crisis regime.

We believe that NATO and other states that adhere to the values of freedom, democracy and human rights must promote the following tactics:

1. to sign an agreement undertaking of not using nuclear weapons first;
2. to state firmly that in case of necessity, i.e. if a non-NATO state applies nuclear weapons against a NATO member state/s, the response will be appropriate, without hesitation and without compromise;
3. the greatest nuclear danger currently comes from the possibility of the application of tactical atomic weapons by the Russian Federation against Ukraine, and NATO must protect Ukraine from nuclear threat. This is an equation with very high potential risk and repercussions for many people, but it requires clarity in the context of deterring and not allowing further outbreaks.

In conclusion, we emphasize:

Amid the extensive invasion of Ukraine by the Russian Federation, European security has experienced notable transformations, marked by an increased proliferation of threats and a reconfiguration of its architectural framework. The onset of active conflict prompted the European Union and its member states to reassess their security and defence strategies, leading to various measures being undertaken over different timeframes. It is evident that both the European Union and its member states must adopt a more inclusive approach in implementing sectoral policies while simultaneously enhancing their strategic partnership with NATO. Closer cooperation between NATO and the EU is imperative, particularly considering the possibility of a resurgence of an "iron curtain" in Europe.

Within the Euro-Atlantic geopolitical sphere, the eastern flank emerges as particularly vulnerable due to its proximity to the Russian Federation. In response, NATO and its allied states are undertaking significant measures to bolster security and defence. These efforts include consolidating resources, increasing investment in the security and defence sectors as a percentage of Gross Domestic Product (GDP), and deploying military groups under the NATO banner. The Alliance's heightened focus on the eastern flank is entirely warranted, especially considering the geopolitical dynamics surrounding both the Baltic Sea region and, notably, the extended Black Sea basin. Given the latter's adjacency to areas of heightened geopolitical tension, such attention is crucial. The Republic of Moldova, recognized as a neutral state by NATO, receives assistance and support from the Alliance. However, it remains vulnerable in terms of security and defence, as its neutrality offers protection only so long as it remains unchallenged by external threats.

The resurgence of rhetoric centered on intimidation and blackmail, particularly concerning the potential use of nuclear weapons, vividly underscores a crisis in deterrence. While such statements warrant serious consideration, they must also be approached with caution and not exaggerated. However, the complexity of the situation is compounded by various factors, including the resurgence of an arms race, the proliferation of nuclear weapons, and the growing number of states possessing such armaments. The prospect of tactical nuclear weapons being utilized as operational tools in conventional conflicts cannot be dismissed.

However, the deployment of strategic nuclear weapons poses existential threats, with potentially catastrophic consequences for humanity. It is evident that the period of conflict in international relations will endure for a considerable duration, signaling the impending onset of a second iteration of the Cold War.

References

1. Brzezinski Z (2010) Marea tabla de sah—Suprematia americana si imperativele sale geostrategice. Ed. Univers Enciclopedic, București
2. Friedman G (2009) Următorii 100 de ani.Previziuni pentru secolul XXI. Litera, București
3. Juc V (2020) Dimensiuni ale relațiilor internaționale contemporane. FEP Tipografia Centrală, Chişinău
4. Fukuyama F (1997) Sfârşitul istoriei şi ultimul Om. Editura universitară, București
5. The Common Security and Defence Policy. https://www.eeas.europa.eu/eeas/commonsecurity-and-defence-policy_en (visited on 29.03.2024)
6. O Busolă strategică pentru consolidarea securității și apărării UE în următorul deceniu. https://www.consilium.europa.eu/ro/press/press-releases/2022/03/21/a-strategiccompass-for-a-stronger-eu-security-and-defence-in-the-next-decade/ (visited on 29.03.2024)
7. Acordul Berlin Plus. https://www.consilium.europa.eu/ro/policies/defencesecurity/defence-security-timeline/ (visited on 30.03.2024)

8. Cooperarea UE-NATO: Consiliul a adoptat concluzii privind punerea în aplicare a declarației comune. https://www.consilium.europa.eu/ro/press/pressreleases/2016/12/06/eu-nato-joint-declaration/ (visited on 31.03.2024)
9. Declarație comună privind cooperarea UE-NATO, 10 ianuarie 2023. https://www.consilium.europa.eu/ro/press/press-releases/2023/01/10/eu-nato-jointdeclaration-10-january-2023/ (visited on 01.04.2024)
10. Outcome of the Madrid NATO Summit, June 2022. https://www.europarl.europa.eu/RegData/etudes/ATAG/2022/733604/EPRS_ATA(2022)733604_EN.pdf (visited on 01.04.2024)
11. Another budget for a country at war: military expenditure in Russia's federal budget for 2024 and beyond. https://www.sipri.org/sites/default/files/2023-12/sipriinsights_2312_11_russian_milex_for_2024_0.pdf (visited on 01.04.2024)

Hybrid Threats for Republic of Moldova Generated by the Russian Military Invasion in Ukraine

Natalia Albu

Abstract Hybrid threats, blending traditional military strategies with non-traditional methods such as cyber-attacks and disinformation campaigns, have become increasingly prevalent in today's complex geopolitical landscape. The Republic of Moldova, frequently targeted by such tactics, faces significant vulnerabilities exacerbated by Russian disinformation efforts. This paper explores the multifaceted nature of hybrid threats and their impact on Moldova's national security, focusing on the role of disinformation as a tool of hybrid warfare. Russian propaganda narratives, portraying Moldova as a failed project and undermining its sovereignty and democratic aspirations, pose significant challenges. Geopolitical tensions and regional dynamics further complicate Moldova's resilience efforts, intensifying the spread of disinformation and fostering anti-EU sentiments. As Moldova braces for the challenges of 2024, including EU accession negotiations and presidential elections, the threat of hybrid warfare looms large. Efforts to combat disinformation include legislative and institutional reforms, such as the National Security Strategy and the Law on Cybersecurity, aimed at enhancing coordination and resilience. External support from the EU and other partners strengthens Moldova's capacity to counter hybrid threats, emphasizing media literacy, cybersecurity, and strategic communication. In conclusion, Moldova must bolster its resilience against hybrid threats by strengthening legislative and institutional frameworks, enhancing strategic communication, and promoting media literacy. International cooperation and support are crucial in developing effective strategies to mitigate the impact of hybrid threats and safeguard Moldova's sovereignty and democratic aspirations amidst a complex and evolving geopolitical landscape.

N. Albu (✉)
Alexandru cel Bun Military Academy of Armed Forces, Chişinău, Republic of Moldova
e-mail: albunatali@gmail.com

F.-C. Cirstoiu et al., *Medical Response Strategy in Case of Radiation Emergency Caused by the War in Ukraine*, NATO Science for Peace and Security Series B: Physics and Biophysics, https://doi.org/10.1007/978-94-024-2266-5_3

1 Introduction

Hybrid threats have become increasingly prevalent in today's complex geopolitical landscape due to several factors. One key aspect is the evolution of warfare tactics, where adversaries combine traditional military strategies with non-traditional methods such as cyber-attacks, disinformation campaigns, economic coercion, and proxy warfare. These tactics are often employed in a coordinated manner to exploit vulnerabilities in target countries or regions, because the ultimate established goal does not necessarily mean a direct attack on the state, but rather to generate instability on all dimensions like political, military, economic, social, justice, media, etc. As a result, the state becomes vulnerable in relation to internal threats including those from outside, being unable to ensure its long-term national security.

Moreover, the use of hybrid tactics allows aggressors to achieve their strategic objectives while minimizing the risk of direct military confrontation or escalation. By operating below the threshold of traditional warfare, they can exploit legal ambiguities and exploit the interconnected nature of the global economy and information networks to achieve their goals. This is why some authors argue that rather than seeking a consensus in defining hybrid threats, it might be better to analyze when state vulnerabilities tend to appear and determine how to combat them. Nevertheless, it is difficult to confront threats if we do not know what these phenomena are or anything about their configurations.

Furthermore, the rise of hybrid threats is fueled by the changing nature of conflicts in the twenty-first century. Unlike conventional wars between nation-states, hybrid threats often involve a diverse range of actors, including state-sponsored entities, non-state actors, and even individuals or criminal organizations. This complexity blurs the lines between traditional notions of war and peace, making it challenging for policymakers and defense officials to formulate effective responses.

In this context, the Republic of Moldova face a series of vulnerabilities, being often the target of hybrid threats through different soft-power tools. Thus, in the context of hybrid threats, these vulnerabilities are successfully explored or aggravated and become more difficult to neutralize, special when war has returned to Europe. The small country, which is now engaged in the EU accession process, has faced Russian disinformation for over three decades since its independence. But Moldova has a long history of grappling with Russian disinformation since gaining independence from the Soviet Union. This includes propaganda efforts aimed at influencing public opinion, destabilizing governance, and hindering Moldova's efforts to integrate with Western institutions. However, the current levels of disinformation pose a significant threat to Moldova's national security, as they extend beyond its internal political dynamics. *Disinformation*, in combination with influence operations, also plays an important role within the concept of hybrid warfare; both from a threat–and own resilience perspective. In this context, we will highlight some tools used by the Russian Federation to undermine national security and analyze the most widespread narratives of disinformation.

2 Disinformation as Hybrid Warfare and Main Russian Disinformation Narratives in Moldova

Disinformation is a significant element in concepts like Cognitive Warfare, Hybrid Threats and Foreign Information Manipulation and Interference. The common denominator of these three concepts is that adversarial actors use them to harm democratic societies while relying on non-kinetic means to remain under the threshold of war.

Thus, *disinformation* and its functionality have become one of the most important threats which exacerbate political and social polarization, weakening the cohesion and resilience of nations. The World Economic Forum's 2024 report identified misinformation as the top global risk in the short term (2 years), while extreme weather events are the top global risk in the long term (10 years). Additionally, in the long term (10 years), disinformation has been classified as a global risk, securing the 5th position, with climate change-related risks taking precedence.

As the National Security Strategy of Republic of Moldova, adopted in December 2023, points out, disinformation is also an important security vulnerability for Moldova. First of all, there is an excessive concentration of assets in the media market within the informational domain. At the same time, society is exposed, over a long period, to disinformation spread by aggressive actors through both traditional and digital mass media. Additionally, we cannot overlook the lack of security culture and the low level of media literacy among citizens, as well as the weak institutional capacities to combat disinformation.

The 2018 Disinformation Resilience Index ranked Moldova as the weakest performing country (out of 14 states) in terms of *Population Exposure and the susceptibility to Kremlin-led media* (3.2), and *Quality of systemic responses* (3.0). Moldova was attributed the second weakest ranking (after Belarus) in *Digital warfare vulnerability* (2.6). A report published by the Romanian Centre for European Policies (CRPE) in July 2020 shows that Moldova is one of the most affected European countries by disinformation campaigns. In this respect, Moldova is a barometer for the effect of disinformation. Russian propaganda represents the main challenge for the information space in Moldova. In this context, the disinformation is a part of the hybrid war in the Republic of Moldova. More, the spread of fake news has increased in recent years—and with the Covid pandemic and the Russian war of aggression in Ukraine, it has reached new dimensions.

The Russian description of contemporary warfare is based on multi-domain operations aiming to undermine adversaries' ability and will to resist. In this framework, Russia's imperial tendencies are expressed not only through cinematographic productions, mass media, literature, but also in its security and defense strategies, including military ones. For example, integrating the component of *preemptive defense* against the near abroad in its military doctrine after the annexation of Crimea.

On one hand, in the context of the information war triggered by the Russian Federation's illegal annexation of Crimea in 2014, disinformation has penetrated the

Moldovan national information space through Russian TV stations, such as Russia's Perviy Kanal, or Channel One, in Moldova, NTV Moldova or RTR Moldova, that are used by a large section of the population in Moldova. In this context, some of them were suspended for example in October 2023 by the decision of the Commission for Exceptional Situations on the recommendation of the Intelligence and Security Services, because there was evidence that 6 TV stations were subordinated to criminal groups, such as Plahotniuc and Şor, who joined their efforts to destabilize the country. At the same time, 31 web portals will be blocked, 21 of which are allegedly managed directly by Russia.

On the other hand, the long period of domestic political struggle in Moldova and the political control over most local media outlets has been used to discredit political opponents. Plus, in Moldova, but also in Romania, the activity of Russia's *fifth column* has been reactivated. For example, invited to Ciutacu's show (RTV Channel), M. Tauber (indicted in the case of illegal financing of the Shor Party) launched for the Romanian audience the main narratives of the Russian propaganda: „the Republic of Moldova should not stay away from any kind of alliance with the West" (the false solution of disengagement from the regional context) or „Maia Sandu wants to drag the Republic of Moldova into war with Russia".

Actually, intelligence services use the term *fifth column* to define a group of people who undermine and sabotage a nation or group from within, usually in favor of another nation or group. Their activities often include sabotage, disinformation and propaganda, and espionage. Currently, it focuses on online and cyber activities.

In this context, it should be noted that for the first time, an official strategic document recognizes that the main threat to Moldova's national security is the Russian Federation, reflecting the fact that *the Russian Federation will not soon give up its hostile actions against Moldova: therefore, we must learn to live under the conditions of a long-lasting and high-intensity hybrid war*. The NSS thus marks a paradigm shift in (–) accepting that Moldova faces threats it cannot cope with and (–) adapting to the new realities of the security environment by developing national resilience.

It is not new that Russia is waging a full-scale hybrid war against Moldova, including using energy weapons, cyberattacks, disinformation, and influence operations. Hybrid threats constitute a combination of different types of tools, some expected and known, some unexpected and clandestine, applied to achieve an undeclared strategic objective, and without officially admitting to doing so.

Russia has traditionally promoted various manipulated history-based narratives that distort reality. Its mass media dominates Transnistria, the Gagauzia autonomous region, as well as targeting other ethnic minorities. At the societal level, the Moldovan Orthodox Church also plays an important role by the trust of the population, an extremely conservative part of the Russian Orthodox Patriarchy—a substantial source of the *soft power* in both inside the country and in the so-called *near abroad* (The term refers to the independent republics that emerged after the dissolution of the Soviet Union). In regions of the country dominated by the Russian-speaking population, the support of the clergy for pro-Russian politicians is particularly high. The selfishness of Russian Orthodox culture; the status of a state in the *near abroad* and the demonization of Europe, all demonstrate that Moscow behaves like a contemporary Tatar-Mongol empire. From a geopolitical perspective, Moscow uses history-based "messianic" projects to promote an expansionist and neo-imperial policy.

At the same, from a strategic perspective, the new Foreign Policy Concept (2023) underscores Russia's desire to assert influence and project power in its area of interest *near abroad*, leveraging historical and cultural narratives to justify its actions (false historical narratives for justification the illegal annexations and invasion). By framing itself as a *civilized state*, Russia seeks to legitimize its involvement in shaping the cultural and civilizational landscape of the Wide Black Sea area, potentially impacting security dynamics and regional geopolitics. Thus, promoting the false narrative that *'Moldova's place is not in the Western world'* through the lens of the following false messages: (–) Independence, sovereignty and democracy of Moldova—a failed project; (–) Moldovans' mentality is backward; (–) Moldova is ruled by foreigners (a. e. puppet of the West/USA); Moldova is being drawn into the war; (–) Moldova is doomed to poverty.

Looking beyond Moldova's geopolitical context, the resilience to hybrid threats may be viewed through the level of socio-political cohesion, and according to national studies, Moldova must increase its level of social cohesion based on four core elements: legitimacy, acceptance, participation, and affiliation. The dimension *acceptance* (the level of solidarity among citizens) has a higher score than average value in the context of the refugee's crises.

At the same time, the situation of Ukrainian refugees has become a narrative exploited by proxy actors. For example, the most widespread false narratives from military invasions in Ukraine, were identified with reference to „Ukrainian refugees pose a threat to public security" or „Children of Ukrainian refugees attending educational institutions in Moldova are aggressive and teachers complain about them". According Rapport *One Year Later: Online Discourse in Moldova about Ukrainian Refugees* elaborated by Internews in collaboration with the International Organization for Migration, the main Russian narratives shared by Moldovian social media aim to create social tensions between the refugee and host communities. These narratives exploit both the socio-economic ramifications of the war in Ukraine and the domestic political competition along geopolitical lines, while capitalizing on the considerably polarized political and media environment.

Ethnic and cultural groups with a history close to Russia are also more exposed to these messages, such as ethnic Gagauz and Russians in Moldova. The majority of high-risk data was collected on Telegram (67%), following by Facebook (22%) and Viber (11%). According national survey (May, 2023), Telegram has emerged as a popular platform for right wing and extremist groups. Also, Russian language websites may be more likely to spread such discourse.

Social cohesion is also affected by the Transnistrian problem. In addition, the impossibility of controlling the districts on the left bank of the Dniester affects human rights and fundamental freedoms in the region (e.g., access to justice, freedom of movement, freedom of expression) and limits the possibility of ensuring human security for all Moldovan citizens stemming from cross-border or asymmetric threats. At the same time, certain challenges may arise that test your ability to respond to certain crisis situations. So, beyond false bomb alerts, in June 2023, Tiraspol accused the Chisinau authorities of blocking the delivery to Transnistria of a part needed to repair a tomograph of the hospital for veterans in the region.

In fact, Tiraspol had no real premises to accuse Chisinau, because previously, through the dialogue channel of political representatives, Tiraspol was informed about the need to initiate the authorization procedure with the competent national authorities of the requesting medical center, equipment of this kind can be managed in compliance with radiological safety rules.

However, the Tiraspol authorities continue to restrict, without justification or using exceptional situations as an excuse, access to the Transnistrian region. One of the most common practices of the de facto administration that unduly restricts the freedom of movement is the declaration of undesirable persons or organization and the prohibition of access to the Transnistrian region for a certain period of time. Plus, according to the 2023 Freedom in the World report classification, the unrecognised region continues to be rated as *not-free*.

Moreover, the Moscow authorities, in the last 2 years have promoted the false narrative that the Moldovan authorities are responsible for the stagnation in the conflict settlement. At the same time, the Russian narrative misconstrued the military upgrade of the Moldovan army as a process of dismantling constitutional neutrality on the way to NATO membership and an alleged militarisation posing threats to the Transnistrian region.

Thus, specific to the context of the war in the neighboring country, the concept of *neutrality* became discussed from several perspectives: as necessity versus incapacity, at the same time being a tool of propaganda and disinformation on the part of Moscow's proxy actors.

In this framework, the Moldovan society, media and state institutions face two major problems—internal and external disinformation. The level of internal disinformation has increased in intensity. Domestic propaganda is disseminated by various actors, including politicians, politically affiliated media sources, institutions with pro-Kremlin editorial policies, and/or Eurosceptics. A tool for spreading disinformation was and still is the Telegram messaging app or TikTok. On this platform, for example, messages were spreading that incited hatred and claimed that Ukrainian refugees from Moldova. Regarding foreign disinformation, it generally focuses on anti-Western messages targeting entities such as the EU, NATO and the US.

A still current propaganda refers to LGBT. Russian narratives promote the view that if Moldova follows the path of liberal EU policy, then this way will automatically turn Moldovan children into homosexuals. Another example of misinformation and manipulation of citizens refers to the period before the official start of the electoral campaign for the election of the Bashkan of the Gagauz autonomy, Cimpoes S., a member of the People's Assembly (PA), who visited Moscow on 17 March 2023 and met with the Deputy Chairman of the State Duma, Tolstoy P. During his visit, Cimpoes S. stated, among other things, that the Moldovan authorities had taken decisions contrary to the interests of Gagauzia, citing as an example the Law on Ensuring Equality, which legalised parades of sexual minorities. His claim does not correspond to the provisions of the law, and is therefore false, and the law does not legalise sexual minority parades. In fact, the Law on Ensuring Equality promotes non-discrimination and equality, in accordance with Protocol 12 to the European Convention on Human Rights (signed by Moldova in 2000). In this context, another narrative is promoted such as *Russia is good, Europe is bad*—the soviet propaganda *while nothing good came out of Europe*. Thus, Russian narratives promote a clear dichotomy between Russia, the Soviet Union and the West.

Thus, all the hybrid instruments used by the Russian Federation and proxy actors tend to promote fear among the population and to undermine trust in government. A relevant example is the issue of a possible Zaporozhye accident is elucidated not so much through the prism of who could cause this nuclear disaster, but through the government's lack of ability to create shelters, manipulating public opinion. Russian officials have emphasized that a large-scale nuclear disaster is in the interests of the US and its allies. In fact, Russia has turned nuclear weapons into an offensive tool to influence international agendas and regional politics. Putin's decision to move part of the nuclear weapons to Belarus puts psychological and media pressure on Ukraine's partners, but it does not change the security situation in the region—neither for Ukraine nor for Moldova.

Finally, beyond the internal or external nature of the source of the misinformation, the same narratives can be promoted in different contexts. For example, in the context of anti-EU narratives, those referring to the Transnistrian region are also topical. An MP and journalist from so-called "Transnistria" promoted falsehoods that negotiations for Moldova's accession to the European Union are a *cover operation* for the so-called *annexation* planned by Romania. Or, another issue is about referendum on EU integration, by organizing it, Chişinău renounces Transnistria—main narrative of the Russian Foreign Minister, S. Lavrov. At the same time, this thesis promotes the old narrative on the disintegration of the Republic of Moldova.

3 The Hybrid Challenges of 2024 and the Existing Prevention Framework

Russia's disinformation operations in Moldova will increase this year amid the Kremlin's desire to do everything in its power to derail Chisinau from its European course. The aforementioned narratives represent manipulation and propaganda efforts aimed at fuelling tensions and undermining efforts to bring Moldova closer to the European Union. This year is a crucial one for Moldova, which will open accession negotiations with the EU and at the same time have presidential elections this autumn and, most likely, a referendum on European integration. This year is a crucial one for the Republic of Moldova, which will open negotiations on accession chapters with the EU and at the same time will hold elections this autumn.

Specific to 2024 is that it is planned to organise a referendum on EU accession on the same day as the elections. Therefore, the referendum will most likely take place jointly with the presidential election. The authorities' plan to organize the referendum is contested by the opposition, which believes that it was designed to help Maia Sandu secure a new mandate and quickly became the target of propaganda and disinformation campaigns. In recent months, Russian propaganda has also intensified a media campaign about Moldova allegedly making preparations with the support (or, in different versions of the narrative, at the behest) of the West and/or Ukraine for an attack on the separatist region of Transnistria.

Thus, on 5 March this year, the Head of the Information and Security Service, A. Musteață, said at a press conference that the SIS had come into possession of data indicating an unprecedented level of intensity of actions by the Russian Federation. The aim of these actions is to compromise Moldova's accession to the European Union, to subvert democratic processes and to keep Moldova under Russian influence.

Also, there are premises of destabilization intentions in ATU Gagauzia. The head of the SIS said that the leadership of ATU Gagauzia is considering possible more radical actions in relation to Chisinau by initiating actions that would destabilise the security situation in the autonomy. In the context of these issues, it is clear that there is a hard pro-Russian and anti-European core in southern Moldova, and a new stage of confrontation has begun with the election of the next Bashkan of Gagauzia in 2023, Evghenia Gutul. Her mandate is characterised by aggressive rhetoric towards Chisinau's domestic and foreign policy, constantly highlighting the differences in geopolitical preferences of the region's population towards the current government. Due to several factors, Gagauzia and Taraclia hves limited integration into Moldovan society, which is reflected in the different attitude towards the European future of the Republic of Moldova. Factors such as the high level of poverty, limited education, low knowledge of the Romanian language and the prevalence of anti-European discourse in the region contribute to a limited understanding by the population of the potential benefits of EU integration.

The subversive influence of the Kremlin's agents is exerted including through proxy parties, created with the aim of destabilising the political situation in the country and undermining confidence in the pro-European ruling party. In addition to political parties that have openly declared their pro-Russian preferences, we are seeing a growing manifestation of other political actors, who although positioning themselves as pro-European, have had links in the past with pro-Russian parties and Kremlin power agents. The current mayor of the capital Ion Ceban is the most prominent example of this, who protested against the Association Agreement between Moldova and the EU in 2014, and 10 years later declares that „European integration should become a national idea".

There are also intentions to provoke a political crisis in the Republic of Moldova manifested by actions to identify and possibly corrupt a number of politicians and deputies, local councilors, representatives of pro-European parties, which would cause disillusionment of citizens in the political class and in the EU integration of the Republic of Moldova. Moreover, there is the intention of the Russian Federation to finance manipulative studies on social and other issues with the aim of influencing the political preferences of Moldovan citizens, i.e. to manipulate public opinion.

Moscow will throw all its disinformation weapons into the fray and will provide considerable sums of money through its proxies, but also directly, to achieve its objective. Artificial intelligence will be one of the methods used in particular.

Actually, the disruptive capabilities of manipulated information are rapidly accelerating, as open access to increasingly sophisticated technologies proliferates and trust in information and institutions deteriorates. Recent technological advances have enhanced the volume, reach and efficacy of falsified information, with flows more difficult to track, attribute and control. For example, Ilan Șor (fugitive oligarch, sentenced to 15 years in prison in Moldova) successfully uses artificial intelligence to promote his blog online, with around 100 portals managed from Asia, when his pages are legally blocked. Another example in our country is the use of artificial intelligence to create *deepfake videos* used to manipulate information or create fake situations, including the image of President Maia Sandu was used to create fake videos, presented on social media as news broadcast on TV channels. At the international level, there is talk of developing legal instruments for governing technologies, but the international community is very slow in adopting new rules and institutions to deal with the new challenges, and dilemmas over issues of national sovereignty and democratic legitimacy persist. In this context, significant international public engagement is badly needed, as national governments and companies cannot solve these dilemmas alone. The Republic of Moldova does not yet have a legislation allowing the sanctioning of forgeries, even more so of deep fakes.

Thus, we cannot neglect the complexity of the hybrid threats to Moldova's national security, and the response requires complex actions, but also innovative approaches, where resilience becomes an increasingly topical issue.

Indeed, the overall level of resilience to disinformation in Moldova is considered to have changed only since 2018 and remains quite limited, despite several legal and institutional changes. In that year, a bill banning the rebroadcasting of Russian content was adopted, but it was repealed in 2020. In this framework it is the so-called Anti-Propaganda Bill. Informally, it was directly aimed at Russia, which was not among the ECTT signatories, but the pro-Russian governing alliance revoked it in 2020.

Also, during that period, the Parliament adopted an Information Security Strategy (ISS) and a new Audiovisual Media Services Code. Both documents are not perfect, plus their implementation has been slow and insufficient. The ISS has four pillars: (1) cyber security, (2) media and information security, (3) operational capacity building, (4) internal coordination and international cooperation.

However, it is considered that information security and cyber security are two very different things and should be treated separately, not grouped in the same strategy. With the military aggression of the Russian Federation in Ukraine, there is an awareness of the importance of protecting the information space of the

Republic of Moldova. In this regard, it is important to mention that among the first measures was the adoption of the decision of the Commission for Exceptional Situations of 2 March 2022, which suspended the transmission of informational and analytical programs produced in the Russian Federation. In order not to create confusion and other premises for misinformation, Russia's non-ratification of the European Convention on Transfrontier Television was used as a legal criterion. And in June 2022, the draft of the *Law on the modification of the Code of Audiovisual Media Services* of the Republic of Moldova No. 174/2018 was approved. Importantly, the introduction of the concept of disinformation is inspired by the definition in the European Union Code of Best Practice on Disinformation, plus the changes introduced in that law were extended to online platforms. However, combating disinformation in the online space has become the most problematic.

Recently, a new policy document was adopted—the National Mass Media Development Program for the years 2023–2026 and the Action Plan for its implementation. The document was developed on the basis of the National Media Development Concept aims to create an effective regulatory framework for the activity in this field. The aim of the Program is to implement coherent public policies in the field of mass media, which would strengthen the role of the media system in ensuring the constitutional right to information and in strengthening a democratic society.

To protect the integrity of its democratic processes and to preserve its Europeation integration, democratic course, the Republic of Moldova has recently taken concrete legislative and institutional steps in strengthening national and state security. An important step towards updating the national security system to current conditions was the adoption in December 2023 of a new National Security Strategy containing valid provisions in the new regional and international security architecture. The strategy pursues 3 key objectives: protecting the security of citizens, building a resilient state and joining the European Union. It is also important that the adoption of the new Strategy raises awareness of the need to strengthen, develop and modernise the national defence system in order to guarantee the sovereignty and integrity of the country, and increasing national resilience to hybrid threats is a valuable factor in strengthening the security culture. And from a sectoral perspective, cybersecurity is another basic pillar for strengthening Moldova's security and defence sector, especially since the *Law on Cybersecurity of the Republic of Moldova* was adopted with EU support on 2 May 2023. The law aims to improve resilience to hybrid threats and increase the cyber resilience of public sector organisations and critical infrastructure in the country.

If we refer to institutional tools in the fight against disinformation, the main public authorities with responsibilities for monitoring, preventing and sanctioning disinformation actions are the Broadcasting Council and the Intelligence and Security Service. Specifically, the Broadcasting Council holds direct responsibility for monitoring and penalizing TV stations within the Republic of Moldova for engaging in activities related to disinformation and propaganda, in alignment with the regulations outlined in the Audiovisual Media Services Code. On the other hand, the Intelligence and Security Service is responsible for monitoring the activities of other stakeholders involved in exercising freedom of expression, notably electronic press and social networks.

It is also important that in order to fight effectively against disinformation, the Centre for Strategic Communication and Countering Disinformation has been established. The Centre's mission is to strengthen cooperation between institutions in the fight against disinformation, manipulation of information and foreign interference, which threaten or may threaten national security and jeopardise the achievement of national interests. The institution will implement the necessary measures to ensure the security of the information space and to increase the resilience of society to threats.

At the same time, we cannot neglect external support, such as that of the European Union in building capacity to combat hybrid threats. For exemple, the EU Delegation in Chisinau manages six strategic communication and media literacy project programmes funded by the FPI, under the Supporting Independent Media and Information Resilience (SIMIR) in Moldova programme implemented by Internews Moldova (for ex., Internetu Graieste—online broadcast, disinformation debunking/regional approach, Community Plus—online broadcast, disinformation debunking, ADEPT—awareness raising/studies in Gagauzia, Youth Media Centre—video/demonstration of conspiracy theories, Watchdog.md—4 surveys on perception of disinformation and main narratives). Through the same project, the Audiovisual Council received support to improve its monitoring capacities and 17 independent media institutions received grants.

At the same time, the European Union Partnership Mission to Moldova (EUPM Moldova) was launched last year as part of the EU's Common Security and Defence Policy. The objective of this civilian mission is to enhance the resilience of the country's security sector in the area of crisis management and hybrid threats, including cyber security and countering foreign intelligence manipulation and interference. The mission will provide strategic advice on strategy and policy development and identify needs to strengthen capabilities for early warning, detection, identification, threat attribution and response to hybrid threats. EUPM Moldova will also implement projects in these areas. Hence the expectations of the Moldovan authorities to develop concrete steps to strengthen cybersecurity with EU assistance.

Also noteworthy is the work of the EU Security Hub for Internal Security and Border Management of the Republic of Moldova, which hosts regular meetings between EU representatives, European agencies, EU Member States and Moldovan authorities to support cooperation in the field of European security and border management. The Hub contributes to increasing Moldova's security by mitigating risks in such areas as illegal trafficking of arms and prohibited substances, illegal migration and cybercrime.

We also cannot neglect external assistance in related areas to combat disinformation and contribute to the country's resilience. Thus, noted that Sweden support SIS in building up its StratCom capacities to respond to hybrid assaults, communicate positively with the public and to position its successes more focused to pro-Russian population parts. Moreover, expert support to identify gaps, tools, better workflows will be provided. Following a baseline assessment, the project will procure needs-based OSINT tools.

USAID through the implementation of projects to strengthen societal cohesion supports resilience, stability and democracy building initiatives in the Republic of Moldova. And the main projects that relate to strengthening the rule of law, anti-corruption mechanisms, strengthening integrity in the security and defence sector are supported by USA, NATO, UE.

Last but not least, responses by independent media and civil society to disinformation and other forms of hybrid threats play a special role. Thus, several NGOs actively contribute to issues such as countering disinformation and increasing security culture, including the Independent Press Association, the Center for Independent Journalism, the Center for Investigative Journalism, the Watchdog Society, the Platform for Security and Defense Initiatives, the Foreign Policy Association, etc. Several websites have also been set up to expose and attempt to combat disinformation—www.stopfals.md, www.mediacritica.md.

These initiatives are appreciated, but they have to compete with the well-funded hybrid tools of the Russian Federation that rely on trolls, sponsored advertising on social networks, optimised search engines, the use of artificial intelligence and attractive click-bait headlines that try to go viral. This calls for a synchronised and coordinated approach to increase the impact of these initiatives, but also to increasing security culture by developing critical analysis. Moreover, in the context of the hybrid warfare being waged both nationally and regionally, the year 2024 will be characterised by a tendency to manipulate public opinion. The latest polls show that large sections of the population lack confidence in the state's ability to provide security, plus poor communication continues in the security and defence field, and public and political support for decisive measures to strengthen them remains dramatically low. This is driven by the low level of the security culture. Therefore, in the public space, there are two main trends of attempting to develop a culture of security and resilience in the face of disinformation campaigns. In this context, the involvement of civil society, academia and the media is vital.

4 Conclusion

Indeed, the Russian Federation has used various subversive methods to intervene in the political processes in Moldova and to influence the country's foreign policy vector. These methods have ranged from manipulation in the energy sector and economic pressure to the funding of extensive propaganda campaigns and the infiltration of the so-called fifth column into various institutions and political parties with the aim of undermining democratic processes and overthrowing the constitutional order. The situation has become even more worrying in the context of the Russian invasion of Ukraine, as the threats to Moldova's national security have also developed a military character. During the 2 years of war, the airspace of the Republic of Moldova has been repeatedly overflown by Russian missiles, while

Moscow has repeatedly threatened that it intends to establish a connection with the Transnistrian region if it succeeds in conquering the south-western regions of Ukraine. Although this scenario is unlikely at present, the most pressing threats emanating from Russia remain hybrid threats, particularly in terms of infiltration of the country's political and electoral processes by Kremlin agents.

In this turbulent context, the Republic of Moldova continues to have many weaknesses in its resilience against disinformation. Although it has taken some tough measures against the classic disinformation channels of TV, radio and some websites, Moldova still has its hands tied in legislating social media activity, where Russia has moved its anti-European propaganda narrative. The authorities' hesitations are linked to the perception that they could censor free expression, which gives Russia fertile ground to step up its propaganda on social media. Russian propaganda is particularly rampant on Telegram and TikTok, which have become alternative pseudo-sources of information but easy entertainment content. Control over the messages that feed the media space, and indirectly the collective thinking in society, is a critical element of national security, without which the viability of a state is under threat from external interference.

Often, narratives serving Russian interests exploit both the socio-economic ramifications of the war in Ukraine and the internal competition between Moldovan political actors based on the country's external orientation. The geopolitical nature of the political struggles is aggravated by the considerable degree of polarisation of society. Also, media experts and researchers of the disinformation phenomenon are of the opinion that such manipulations are not sufficiently punished and thus indirectly encouraged to be amplified on social networks.

At the same time, Moldova's national security is deeply affected by the war in the neighbouring country, which raises concerns about the state's resilience and its ability to cope with hybrid threats. For these reasons, it is important for the state authorities to focus on certain directions of action:

- Strengthening the institutions of law in the context of the EU accession process;
- Ensuring the effective functioning capacity of the Centre for Strategic Communication and Countering Disinformation, in particular with human, financial and technical resources in order to facilitate proactive communication on the main risks generated by disinformation and manipulation of public opinion and to protect the information space of the Republic of Moldova, in particular to ensure a rapid response and effective functioning of strategic communication;
- Facilitate cooperation between state institutions and the non-governmental sector in promoting initiatives to educate society about the risks of misinformation in the online environment and promote media literacy in society;
- The need for regulatory intervention and the institutionalisation of a concrete mechanism to counter misinformation online;
- Strengthening cyber security through investment in infrastructure, cyber security capabilities and promotion of the cyber security among civil servants;
- Targeting according to the seriousness of the facts to counter disinformation activities directed against national security and public order, but the targeting

must include transparency tools to prevent abuses and ensure respect for human rights in case of policy change;
- Develop and maintain a constant dialogue with different regions of the country (especially Russian-speaking localities and ATU Gagauzia);
- Strengthen cooperation with external partners, in particular with the institutions and agencies of the European Union, Western states and NATO, in order to take over best practices in the area of national resilience and improve the capacity to respond to hybrid threats.

Finely, hybrid threats pose a complex and evolving challenge to the international order. As they continue to evolve and adapt, policymakers and security experts must remain vigilant and proactive in developing effective strategies to mitigate their impact and maintain global stability.

Nuclear Geopolitics in the Context of Redefining the Great Strategic Game

Veaceslav Ungureanu

Abstract The geopolitical structure of the post-cold war international system is determined by the arrangements of the world geopolitical architecture of the international security that shows itself as a process where different geopolitical configurations of the international relations alternate by reshaping the spheres of geopolitical influence expressed as some geopolitical interests and values promoted by the state international actors and by carrying out actions and individual or collective interactions by the geopolitical subjects to ensure a sustainable geopolitical structure of the international security. In conditions of the global anarchy of the international relations that determines the world hierarchy of the power, it is a quite natural thing to have a competition between the major actors for the nuclear geopolitical hegemony. This is conditioned by the need to participate in the process of ensuring the international security both alone and through political and military alliances, developing its own nuclear geopolitical management of the international security that determines the geopolitical configuration of the post-cold war international system, ensuring or undermining at the same time the process of guaranteeing international security by further keeping the tendency to enlarge its spheres of geopolitical influence and intensifying the role of military and nuclear factor that has not decisively lost its relevance and actuality, but on the contrary has strengthened.

1 Introduction

The end of the cold war has profoundly shaken the scientific academia world and the circles of specialists in the field of geopolitics, securitology and international relations, depriving them, on the one hand, of topics intensively exploited and scientifically explored in the past, and, on the other hand, making them face new, scarcely known, explained and understood realities, and they were compelled to

---

V. Ungureanu (✉)
State University of Moldova, Chișinău, Republic of Moldova

F.-C. Cirstoiu et al., *Medical Response Strategy in Case of Radiation Emergency Caused by the War in Ukraine*, NATO Science for Peace and Security Series B: Physics and Biophysics, https://doi.org/10.1007/978-94-024-2266-5_4

engage into developing new concepts and solutions necessary for the management and theoretical substantiation of actions and processes to ensure the post-bipolar nuclear security. It is worth clarifying that neither F. Fukuyama, nor S. Huntington have further developed the ideas that rendered them famous. Maintaining the opinion that there is no alternative to liberal economy and democratic norms, the former opted for other research issues, with a predilection of moral nature, while the latter addressed the world geopolitical order from other, mainly realistic, perspectives, one of which being polarity. Therefore, S. Huntington rationalises on the evolution of the world geopolitical order in the post-cold war era, arguing that after a moment of unipolarity, the mankind will go through several decades of uni-multipolarity, while the twenty-first century will feature multipolarity. We notice that S. Huntington, like certain realists, such as K. Waltz, views the unipolarity as a short-term phenomenon and that it will imminently evolve towards a multipolar geopolitical structure, a masterly assertion almost generally accepted in the academic environment [1, p. 45–48].

The behaviour of the great nuclear geopolitical powers and their satellites in the period of the post-cold war has been marked by actions of reoutlining the spheres of geopolitical influence through hard power in order to achieve the geopolitical interests determined by the need to redefine the policies of ensuring the national security. For the first time in history, a non-Eurasian geopolitical power imposed itself not only as the main arbiter in the power geopolitical relations of Eurasia, but also as a world geopolitical superpower. The collapse of the Soviet Union represented the last step in the rapid geopolitical rise of a nuclear power from the Western hemisphere—the United States of America—as the only and, as a matter of fact, the first truly global geopolitical power. The fundamental issue is whether the United States of America, engaged on a planetary scale, through its capacity and manner of exercising the world geopolitical supremacy, by fulfilling its Eurasian geopolitical interests and by increasing its nuclear capacity of ensuring the national security and the security of satellite states, can deal with geopolitical power relations of Eurasia and, especially, whether it can prevent the emergence of an equal dominant Eurasian geopolitical power.

2 Geopolitical Nuclear Balance of the Bipolar International System

St. Hoffmann specified that the bipolar international system that emerged after the Second World war was distinguished by heterogeneity. At the same time, it was characterised by a high degree of flexibility and moderation, resembling, to a certain extent, to the classical type of balance, while not being multipolar. The peculiarity of the contemporary bipolarity is that the balance of nuclear powers, pursuant to R. Aron's estimations, was not the subject of the coalition. Since the satellites gravitate around each of the two superpowers *blocked* by the presence of the thermonuclear weaponry under the conditions of the *balance of terror*, in the opinion of J.-B. Duroselle, the continuity of alliances against hegemony is no longer registered. Peace is ensured by the nuclear revolution, defined by B. Badie as the capacity of the state in possession of the *absolute weapon* to destroy the enemy before defeating its armies. Also, the order at the global level is maintained by the strategy of nuclear deterrence, that is, the threat of an unbearable but safe counterattack in case of aggression.

The nuclear deterrence was reflected in numerous versions, such as the *stable deterrence,* as described by Th. Schelling and G. Snyder, or the *existential deterrence* developed by M. Bundy. Concomitantly, it is no less true that several doctrines have been presented on the possibilities to emerge victorious in a nuclear war, such as the doctrine described by A. Wohlstetter. The researcher claims that, technically, the adversary strategic forces can be liquidated before their application. In such conditions, the attacking party is able to reduce the counter hit down to a bearable level, being able to dictate its will. On the contrary, K. Waltz rightly underlines that the states with nuclear weapons could have stronger motivation to avoid the war, than conventionally armed states [2, p. 91].

In the same context, V. Juc claims that, in the bipolarity conditions, the international situation was characterised by a high degree of centralisation, expressed through the monopolies of applying the violence and the nuclear weaponry, the latter being the determining factor, which did not allow the frontal collision between those two superpowers. The post-bipolar period witnesses the decentralisation of international relations through the emergence or update of certain factors generating conflict [3, p. 5–9].

Limiting himself only to the research of contemporary international relations and being of the opinion that the pace of change in the fundamental rules of their functioning has increased in the twentieth century, J. F. Guilhaudis divided the bipolar international system in three distinct periods: 1945–1962—period of the cold war; 1963–1979—period of relaxation; 1980–1989—period of bipolarism entropy [4, p. 73]. It is to be noted that the inside bipolarity periodisation by the identification of certain similar chronological limits is frequent. In our opinion, the beginning of the third period could be overrun in 1977–1978, starting from the deployment of Soviet modern nuclear missiles in the European part of the country and, as a countermeasure, the call of the West German Chancellor H. Schmidt to deploy American medium range and cruise missiles in certain West European states.

Pursuant to A. Tuzikov, in the post-cold war period, a polycentric structure of the international system was built with pronounced features of geopolitical hegemonic unipolarity, and the states that do not want or do not prove to be able to align themselves with the new model of the world geopolitical order manifest open aspirations to possess nuclear weapons, associated with the guarantee of sovereignty, a situation which can degenerate into the relaunch of the arms race with the arsenal of mass destruction [5, p. 86]. At the same time, the expert believes that the model of the post-bipolar world geopolitical order was created in the context of resizing the world geopolitical architecture and intensifying the processes of globalisation, while, it should be emphasised, his ideas fit, to a large extent, in the logic of building the concept substantiated by H. Kissinger on the conflict between fragmentation and globalisation, marking the international system at the beginning of the twenty-first century.

P. Hassner considers that the high number of power centres is at risk of becoming rather a source of unforeseeable geopolitical events and of uncontrolled nuclear disorder than a restraint, emphasising that the nuclear balance between the superpowers was established in the bipolar geopolitical world. E. Mansfield and J. Ikenberry, especially the latter, also think that the unipolar geopolitical order proves to be the most efficient form of ensuring the international nuclear security and the stability of the international system due to the capacity of the nuclear hegemonic power to prevail over any coalition and intervene in geopolitical conflicts generating instability by inhibiting warlike moods [6, p. 22–24]. In the attempt to identify the peculiarities of the *international bipolar geopolitical system*, St. Hoffmann considers that it is, at the same time, both revolutionary and moderate, invoking the possibility of mutual assured destruction. K. Waltz named this inconsistency a deterring ambiguity of position, but it is no less true that he himself believed that the nuclear weaponry rendered the bipolar world geopolitically moderate, even if, according to H. Mouritzen's appreciations, who made several comments on the Waltz's theory, he considered it important in the context of his own theory, though secondary in relation to the distribution of nuclear capacities.

3 Geopolitical Nuclear Decentralisation in the Post-cold War Conditions

According to H. Kissinger, following the disappearance of the danger from the Union of Soviet Socialist Republics, the Atlantic partnership remains, for the United States of America, the decisive geopolitical pillar of the world order, beyond the strictly determined principles of ensuring the collective defence, characteristic of traditional alliances, the allied states of the North Atlantic Treaty Organisation have created a network of relations having as objective the creation and strengthening of the foundation for the common political future [7, p. 11]. A. Bogaturov considered that the Eurasian priority of the foreign policy of the United States of America proved to be more and more obvious, due to the fact that the Central Eurasia had the role of the main partner in the global American geopolitics, having become, in the twenty-first century, the new world's geopolitical centre, while the role of Europe has proven to be different, a complementary one, while further remaining important [8, p. 34, 35]. However, H. Kissinger and Ju. Kashlev have noticed the geopolitical retractions at the global level, the first one invoking the influence of the growing weight of the Asian economy on the development of the United States of America, in particular, and of the world, in general, while the latter noted the shift of the world's economic, financial and commercial centre to East and Southeast Asia. Therefore, Zb. Brzezinski found the identification of the most important Eurasian geopolitical centres, claiming that the need to protect them represents an aspect of principle in promoting the global geopolitics of the United States of America.

The duration of the world's geopolitical unipolarity in the period of the post-cold war is conditioned by the nuclear geopolitical competitiveness, according to the classification of Zb. Brzezinski, able to „geopolitically counterbalancing", in K. Waltz's opinion, „the severely disbalanced geopolitical power". In this connection, Zb. Brzezinski shows, in classical geopolitical terms, the main manifestation of the geopolitical interest of this hegemony of the unprecedented role played by the United States of America in the entire continental Eurasia, which served as the point of origin of all previous geopolitical actors to world power. Currently, the United States of America is the nuclear geopolitical arbitrator of Eurasia, and no major Eurasian issue can be solved without the participation of the United States of America or against its geopolitical interests. Zb. Brzezinski has identified at least five key geostrategic actors and five geopolitical pivots (two of the latter partially defining themselves as geostrategic actors) on the political map of Eurasia. France, Germany, Russian Federation, China and India are important and active actors, while the United Kingdom, Japan and Indonesia, though being important states, have not received this title. The expert gives Ukraine, Azerbaijan, South Korea, Türkiye and Iran the role of geopolitical pivots of decisive importance [9, p. 53–54], an assumption proven in the case of Ukraine, which, in our opinion, turned out to be a fundamental geopolitical pivot for European and international security systems, as well as for the national security of great powers.

W. Wohlforth also considers that China and the Russian Federation do not have enough means to face the United States of America, an assumption challenged by J. Mearsheimer, in whose eyes the United States of America is not a global nuclear geopolitical hegemony, though it remains the world's main economic power, while two other great powers are present in the international system—China and the Russian Federation, which have important nuclear arsenals, though without being able to reach the level of the American military potential and have limited capacities of projecting the geopolitical power in the world. J. Anderson, on the contrary, addressing the limits of the Sino-Russian geopolitical partnership, mentions that these two countries have clearly expressed their dissatisfaction with some geopolitical aspects of the American foreign policy, though have not created a serious geopolitical coalition against the United States of America, and few researchers believe they are likely to do it in the future [10, p. 68].

In this context, challenging the very possibility of existence of the unipolar geopolitical structure of the international system, K. Waltz has expressed his opinion that the bipolarity was maintained in the first half of the 90 s, albeit in a modified form, because the Russian Federation preserved its nuclear counterattack capabilities, while other great geopolitical nuclear powers have not emerged. In the spirit of the same argument, *i.e.* possession of nuclear weapons by the United States of America and the Russian Federation, which constitutes a determining force for peace, J. Mearsheimer has also invoked the bipolar geopolitical structure in Europe that remained after the end of the cold war, while the North East Asia has seen the creation of a geopolitically balanced multipolar system, with China as the first great relevant power that still does not possess the qualities of a potential geopolitical hegemony. Proposing the degree of cohesion as the tie-breaking criterion, A. Iakovlev distinguishes the monolithic West and the rest of the world, made up of autonomous components, of geopolitical poles of local significance, but which are not well welded by conceiving the supreme priority of common geopolitical interests and organisational forms, even if the Russian Federation, India and China actually and potentially prove to be influential geopolitical factors within the world geopolitical architecture of international security and the geopolitical structure of the international system [11, p. 180–181].

In this context, Zb. Brzezinski emphasised the ensemble of methods and proceedings for the United States of America, with a view to maintaining and promoting its geopolitical interests on the Eurasian continent. The short-term geopolitical interest of the United States of America is to strengthen and maintain the geopolitical pluralism prevailing on the map of Eurasia, realising the priority of actions in order to prevent the emergence of a hostile coalition that could ultimately challenge the supremacy of the United States of America and pose a threat to the national security, considering the real possibility of each separate state to try this. In the medium term, the geopolitical events could evolve gradually, with the United States of America emphasising the appearance of certain geopolitical partnerships with states that have a geostrategic potential and that would contribute to the building of a more cooperative trans-Eurasian security system. Eventually, in the long term, Eurasia could become a world core of geopolitical responsibility. We align to the expert's assertion pointing out that the failure to solve the new global geopolitical issues extending far beyond the borders of nation states and the intensification of geopolitical concerns of great nuclear powers on ensuring the national security, will be able to determine essential transformations of the geopolitical structure of the world's power. At the same time, considering the fluctuating character of political, economic and regional security processes in Europe and Asia, any successful foreign policy of the United States of America must focus on Eurasia as a whole and be guided by a well-defined geopolitical interest [9, p. 51–52].

H. Kissinger, like Zb. Brzezinski, emphasises the potential of eventually great nuclear powers, which further would enable them to challenge the geopolitical interests of the United States of America on Eurasian continent and compete geopolitically worldwide, while underlining a rational and adequate behaviour of the United States of America in relation to those geopolitical forces which are able to undermine the geopolitical structure of the post-cold war international system. H. Kissinger pointed out that the geopolitics of the United States of America towards the Russian Federation must be thought in compliance with the permanent geopolitical interests and not depending on the Russian domestic politics' fluctuations because Russian authorities will permanently have a geopolitical interest in the *immediate neighbourhood*, while the key problem is whether the relations of the Russian Federation with new independent states must be seen as an international issue subject to the rules accepted by the foreign policy and codified in the international law, or as a result of the unilateral decision of the Russian Federation to keep the post-Soviet region in the sphere of its geopolitical influence.

It is to be mentioned though, based on the 2014 events in Ukraine, as well as the interference in the domestic affairs of the post-Soviet states, including the lack of progress in the settlement of *frozen conflicts*, that the American administrations have not solved this dilemma, considering the ex-Soviet area as a segment of the Russian Federation's geopolitical interests. The geopolitical attitude and behaviour of the United States of America in relation to this area changed after 24 February 2022, when the Russian Federation started a war against Ukraine.

We specify that the evolution of the events in Ukraine in 2014 and the military aggression of the Russian Federation against Ukraine in 2022 could not be predicted by any expert, because it was considered that the geopolitics no longer substantiates the vital interests and actions of the great nuclear powers' foreign policy giving priority to the international law rules, which proved to be unfunctional in regulating certain international disputes or regional conflicts having an interpretative character that the major actors are still using today to promote, achieve and protect their own geopolitical interests in various regions of the world [12, p. 26].

Acting from the geopolitical point of view in the ex-Soviet area, without consideration for the established rules and norms of the international law, prioritising the geopolitics at the expense of the legal provisions established in the 1994 Budapest Memorandum, signed and ratified by Ukraine, the United States of America, the United Kingdom and the Russian Federation, which tackles the nuclear disarmament of Ukraine and provides security guarantees for the independence thereof, the Russian Federation confirmed once again the emphasised existence of the geopolitical dimension within the foreign policy of great powers, which is determined by the geopolitical interests outlining the spheres of geopolitical influence, in which, as the Russian Federation shows, it can also act deliberately using military means for its defence.

At the same time, in the aftermath of the Soviet Union collapse, Ukraine, proclaiming its independence, has become the third nuclear power in the world, with nuclear weapons stockpiles greater than the United Kingdom, France and China taken together, while through this treaty, Ukraine gave up Soviet nuclear weapons on its territory, which were transferred to the Russian Federation, and instead, the United States of America, the United Kingdom and the Russian Federation committed to respect Ukraine's independence, sovereignty, territorial integrity and existing borders [13, p. 273–274]. Therefore, according to the Russian Federation, the sphere of the geopolitical influence in the post-Soviet area represents a security area made up of buffer states aimed at counteracting the integrationist geopolitical processes carried out by the North Atlantic Alliance and the European Union, by extending towards East the area of Western values, which represents a threat to the national security of the Russian Federation [12, p. 26].

We consider that in the context of the military aggression of the Russian Federation against Ukraine it is imperatively necessary that the United States of America create, still preserving its superpower status, auspicious geopolitical conditions to strengthen the security guarantees on the European continent not only for the member states of the North Atlantic Treaty Organisation, but to extend them for the non-member states, including those in the post-Soviet geopolitical area.

As for the Russian Federation, the diplomatic, political and economic international isolation will deepen in all aspects and in all international organisations, despite being a great military and nuclear power with global geopolitical ambitions, the role of which should not be underestimated in the world politics, if the Russian Federation does not relinquish its foreign politics by achieving its geopolitical interests using hard power in the post-Soviet area and does not remove animosities with ex-Soviet states, respecting their independence, sovereignty and territorial integrity.

A stage of the post-cold war in which the geopolitics will have a fundamental role in the process of building two „Western and anti-Western" security blocs (alliances) and reconfiguring the spheres of the geopolitical influence the distribution of which will be carried out on the basis of adopting the position and stance officially declared by the state concerning the war in Ukraine, the awareness level of the capacity to ensure the national security and defence under the conditions of a possible invasion from the Russian Federation and uncompromising geopolitical support from the great nuclear powers [12, p. 27].

We argue that under conditions of geopolitical rivalry of these two security blocs, other issues of regional and international security will be addressed, such as: the hostile American-Iranian relations and the Nuclear Agreement with Iran, which must serve as the guarantor of a nuclear stability in the region and will put an end to the development of the nuclear programme of Iran benefiting in exchange for it from a relaxation of sanctions imposed by great powers, in particular, in the economic field; the rivalry in the geopolitical relations between the United States of America and the Russian Federation is generated by the suspension of the participation of both parties in the Intermediate Nuclear Forces Treaty, signed in 1987 by the leaders of both states, which bans the development, production and deployment of ground launched nuclear missiles with a range of 500–5500 km, repeatedly invoked by the Russian Federation in order to challenge the deployment of the anti-missile shield in Romania or Poland, which could be used for launching the nuclear missiles by the United States of America, as well, arguing that the Russian Federation violated the provisions of the Treaty by developing latest generation nuclear missiles, and the fact that China should be party to a new Treaty negotiated by parties, considering that the old Treaty disadvantaged the United States, since China developed intermediate range nuclear missiles facing no constraints that should be regulated at the international level.

The geopolitical American Chinese relations are endangered by an eventual crisis in connection with Taiwan or by the emergence, in China, of an aggressive and hostile regime determined by its domestic political and economic evolution or by the cooling of the relations between the United States of America and China. In this regard, China could become a particularly destabilising force in the region and in the world, resorting to enormous pressures on the United States-Japan relations by generating policies of geopolitical disorientation of Japan, which would compromise the geopolitical stability of Southeast Asia, influencing, as well, the regional geopolitical behaviour of India, an essential country for the stability of South Asia [12, p. 28].

T. Marshall believes that the foreign policy of the United States of America in the twenty-first century should focus on the regions of Asia and the Pacific Ocean, because half of the world's population lives there and, if India is also considered, then these areas will provide, by 2050, half of the global economic production. Also, Japan, Thailand, Vietnam, South Korea, Singapore, Malaysia, Indonesia and other states in the geopolitical region of Southeast Asia promote a foreign policy of intensifying, deepening and diversifying the geopolitical relations of cooperation with the United States of America, being convinced that, if not united, they will fall under the Chinese geopolitical hegemony.

For this reason, the United States of America are going to focus their geopolitical interests establishing themselves in this geopolitical region, bringing financial investments and a massive military presence exercising a tangible influence, paying attention to the limited military actions in order to secure allies in the event of open conflicts. The expert claims that the twenty-first century will see the great powers achieving new geopolitical compromises in the region of the Pacific Ocean. Most compromises, but not all, will be likely obtained by China, a vivid example to which is the declaration by China of a zone of aircrafts identification, compelling other states to inform China prior to entering the air space of a disputed territory, and China won by creating and transforming this zone into a subject of geopolitical dispute, while the United States of America, flying over without notification, have also won by disregarding these conditions, which fact represents an enduring geopolitical game [14, p. 88–89].

The visit of the US President on 10 September 2023 to Hanoi can serve as proof to these assertions, resulting in a bilateral agreement signed by both presidents, which raises the relation between the two countries to the level of strategic partnership, a new step forward for the USA plan to counteract the Chinese geopolitical influence in the region of Southeast Asia, by increasing the cooperation between the USA and Vietnam at the highest level to become a power for prosperity and security, strengthening, at the same time, the interaction between the alliances constituted by the USA, such as: the relation between the Quadrilateral Security Dialogue (the USA, India, Australia and Japan) and AUKUS, a Trilateral Security Partnership (Australia, the United Kingdom and the United States).

Another geopolitical dispute in the region of Southeast Asia between the great nuclear powers, emphasised by T. Marshall, is that China will never give up the artificial islands it is building in the Spratly Islands area of the South China Sea and, when in 2016, a Chinese aircraft landed for the first time on one of the airfields built on them, Vietnam and the Philippines officially protested since both countries claimed this area, while the United States of America declared that Chinese actions threatened the region's geopolitical stability and that it would thoroughly monitor the disputed area involving naval and air patrols to assure the allies that they stood beside them, guarantee the freedom of navigation in international waters and deter China geopolitically from being drawn into a regional military confrontation.

According to the expert, the geopolitical interests of the United States of America in relation to Japan are expressed by assuring Japan that the geopolitical interests of the US and Japan vis-à-vis China are identical and receive guarantees from the

Japanese state that the American base on the Okinawa Island will further stay functional to strengthen the Japanese defense capacities, while, at the same time, limiting the military capacity of Japan to intervene militarily by itself in the Pacific Ocean. Concomitantly, the main satellite states in the geopolitical game played by the United States of America in the Southeast Asia region are Indonesia, Malaysia and Singapore, dominating the Strait of Malacca, a strategic zone through which about 12 million barrels of oil are transported daily to China, whose economy needs more and more raw material, as well as to other countries in the region.

The United States secured a major geostrategic advantage and leverages to exert influence on the regional geopolitical conjuncture. The rivalry between the United States of America and China on the dimension of promoting the nationalism to ensure the unity of the Chinese people and the claim to respect the territorial integrity of China through the recognition of the Island of Taiwan as the component part of China cannot be overlooked. The United States of America have signed a Treaty with Taiwan, according to which, if China invades what they consider to be their twenty-third province, which is unlikely, the United States will go to war, which may happen, in T. Marshall's view, only after the official recognition of Taiwan by the United States or after the declaration of independence of Taiwan from China [14, p. 89–91].

As the China's need for gas and oil increases, being dependent on the import of hydrocarbons, the United States of America downsizes its import quota of energy products due to the drilling for the exploitation of shale gas on large areas of the country, trying to secure not only the energy needs, but equally become an energy importer and exporter in the nearest future. In this regard, T. Marshall points out the reduction of the oil and gas flows from the Persian Gulf region to the United States of America that further maintains its geopolitical interests in the region by keeping the Fifth Fleet in Bahrain and supporting the satellite states, such as the Saudi Arabia, the United Arab Emirates, Kuwait and Qatar to prevent China from strengthening its geopolitical power in the region, after it has a powerful military ocean fleet, or Iran, an actor with geopolitical potential of regional power aspiring to extend its geopolitical influence in the Middle East, by promoting an anti-Semitic politics against the State of Israel, which is an allied state of the United States of America in achieving its regional geopolitical interests [14, p. 91].

Pursuant to T. Marshall, the short-term foreign policy of the United States of America in the Middle East lies in undermining the geopolitical and nuclear potential of Iran to reach an interstate agreement to diminish the multiple controversies and dissensions in the relations between these two countries that have divided them for four decades [14, p. 91]. G. Friedman accepts the assumption of T. Marshall, arguing that Iran is not a significant military or economic force, but it is under the sanctions adopted by the West against the implementation of the uranium enrichment program to obtain nuclear weapons, and generates regional instability [15, p. 128].

Also, T. Marshall specifies that the United States of America, as a nuclear superpower, will stick to its geopolitical interests in Latin America, as well, capitalising on them by keeping the Panama Canal open, keeping under control the tolls through the future Nicaraguan Canal to the Pacific Ocean and by monitoring

Brazil's geopolitical rise, so that its influence does not extend over the Caribbean Sea. From the geopolitical point of view, the United States of America will compete with China in Latin America, but not in Cuba, a country for which the United States of America will remove all obstacles to secure its dominance in the post-communist era. The United States of America is one of the geopolitical actors interested in the natural resources of Africa, closely following the evolution of Islamist groups in the North Africa, while China is the best exploiter of the continent's natural resources [14, p. 91–93].

In T. Marshall's opinion, the territorial confrontation between China and other nuclear power, like India, determines the intensification of the nuclear risk to the international security, which materialises in China's geopolitical aspiration to obtain the Indian province of Arunachal Pradesh, while India claims Aksai Chin, which fact gives rise to a diplomatic tension in the geopolitical relations between these two countries with nuclear potential, each having to treat and ensure the security in this region with great caution. The Indian Tibetan border represents an important geopolitical area for both countries, and the eventual attempt on the part of India to change the border with a view to controlling Tibet and the Tibetan Plateau, a region offering it a strategic advantage to force both the advancement towards the centre of China, and the control over China's three great rivers—the Yellow River, the Yangtze River and the Mekong River, could also bring about a regional military conflict between China and India, both states having nuclear weapons [14, p. 55–59].

Contrary to T. Marshall's view, despite all problems that the Russian Federation is facing in the field of domestic policy of modernising the state, we admit the fact that Russia exercises a constant geopolitical influence worldwide, including in Latin America, together with countries that are in opposition to the foreign policy of the United States of America promoted in that region, such as Venezuela, Nicaragua, Cuba, and also trying to balance the geopolitical influence of the United States of America in the Middle East, sometimes to decrease it, as is the case of Syria, creating alliances with leaderships of Syria, Türkiye and Iran to fight the terrorist organisations. To assure its national security, the Russian Federation is also advancing territorial claims manifested through the achievement of geopolitical interests in the Arctic area, deploying stationary military troops in the region, at the same time, being concerned with Greenland and expressing geopolitical interests in the African continent, as well, and having organised, on 23–24 October 2019, in Sochi, and on 27–28 July 2023 in Sankt-Petersburg, the Russia-Africa Summit to rebuild the cooperation relations with the African states and lower the geopolitical influence of the United States of America, the European Union and China on the African continent, launching a geopolitical struggle.

N. Nartov does not exclude the likelihood of a geopolitical rivalry between the United States of America, on the one hand, and, on the other hand, the eventual coalition including the Russian Federation and China created by concentrating the natural resources and the dynamic development of the economy: these states can build „the single Eurasian geopolitical area", which, by its geographical, economic, demographical and raw materials parameters, will be outside of effective geopolitical competition with any other state [16, p. 285].

L. Kryshtanovich shares the same idea, arguing that the concentration of geopolitical potentials of the Russian Federation and China will annihilate all anti-Russian and anti-Chinese geopolitical powers on the world stage, a coalition that all geopolitical rivals will fear. V. Petrov is much sharper, by requesting that the Russian Federation methodically, step by step, create its „geopolitical space", called Eurasia [17, p. 176], which, in our opinion, has already begun by achieving its geopolitical interests in the post-Soviet geopolitical space by applying the „hard power", triggering the active conventional phase of the hybrid war against Ukraine, which may eventually stretch to the states of Eastern Europe, despite the fact that they are members of the North Atlantic Treaty Organisation.

We reiterate that the authorities of the Russian Federation could not but aspire to capitalise on and implement this kind of geopolitical ideas, which, otherwise, do not stand out by novelty, always being present in the political arena: BRICS (Brazil, Russia, India, China and South Africa Organisation), SCO (Shanghai Cooperation Organisation), or the Customs Union (Russia-Belarus-Kazakhstan), rebranded into the Eurasian Union. We emphasise that the results recorded by the first two international organisations are currently below expectations, due, *inter alia,* to the fact that one of the cardinal goals was not „for" geopolitical cooperation, but „against" the American geopolitical hegemony, while the third one has not stood out and runs the risk to repeat the fate of the Council for Mutual Economic Aid, albeit in a narrower format and a much higher degree of asymmetry.

T. Marshall reckons that the Russian Federation shows interest in the regional geopolitical status of demilitarised neutral state with guarantees of non-alignment to the North Atlantic Treaty Organisation for Ukraine, forcing it to accept these conditions by starting the war, advocating, at the same time, for the creation of a buffer zone of neutral states around its borders securing thus its national security, while Ukraine is assigned the central role in defending Russia in case of a possible offensive military attack from the West in the direction of the North European Plain. A pro-Western Ukraine with ambitions of accession to the North Atlantic Treaty Organisation and the European Union, undermining the access of the Russian Federation to the military port of Sevastopol in the Crimean peninsula in the event of diminishing its influence in the Black Sea and the possible location of a naval military base of the North Atlantic Treaty Organisation, is unconceivable by the Russian Federation, prompting it to a direct military aggression against Ukraine [14, p. 29].

It is to be specified that the port of Sevastopol is a strategic point in the Black Sea for the Russian Federation and the only port with warm waters that should secure the access of military ships to all the oceans in the world all the year round, but this is not the case either, because leaving the Black Sea and passing through the Bosporus Strait, the Sea of Marmara, the Dardanelles Strait, the Aegean Sea and entering the Mediterranean Sea is restricted by the Montreux Convention of 1936, by which Turkey, currently a member of the North Atlantic Treaty Organisation, exercises control over this strategic region by monitoring the status of ships, their number and the period of crossing.

At the same time, even after entering the Mediterranean Sea, Russian Federation's ships must cross the Strait of Gibraltar to reach the Atlantic Ocean, and to reach the Indian Ocean they must be allowed to cross the Suez Canal. There is, however, a naval presence of the Russian Federation in the Mediterranean Sea, on Syria's Mediterranean coast, in Tartus, but it is not an important naval force, rather a small naval base tasked with military equipment resupply and logistics assurance of the Russian service men involved in providing military support for the Syrian government [10, p. 39–40].

In the Balkan region, the Western Balkans remain a subregion of geopolitical importance, being at the confluence of the geopolitical interests of four great powers—the United States of America, the European Union, the Russian Federation and China. Each of these geopolitical actors apply available strategies, structures and leverages: the United States—the strategy of Euro-Atlantic security, the European Union—the association and integration through the access to the European institutions and funds, the Russian Federation—energy supply and, in certain cases, the leverage of the Slavic Orthodox community, China—increase of reliance on reimbursable investments through loans to state infrastructure projects. As a matter of fact, two more interregional cooperation institutions are present in the subregion, such as the Organisation for Security and Cooperation in Europe and the North Atlantic Treaty Organisation, and together with the European Union, they aim to contribute to ensuring regional security and stability, including through management of crises in Kosovo and Bosnia and Herzegovina [10, p. 40].

4 Conclusions

The dynamic of redefining the geopolitical interests by the great powers has intensified the resizing of the world geopolitical architecture of the international security that, in turn, has influenced the process of reshaping the post-cold war international system. At the same time, in order to ensure the nuclear geopolitical dimension of the international security, the great powers continue to pursue the achievement of ensuring the geopolitical balance of the international security system both on the account of the major nuclear actors and through some medium and small nuclear international actors, thus enlarging the spheres of geopolitical influence, but avoiding as much as possible their interference that may trigger a nuclear military conflict between the rival geopolitical actors. It is certain that strengthening the resilience of the international security system on the nuclear geopolitical dimension largely depends on the degree of the nuclear powers' involvement into the geopolitical processes of ensuring regional nuclear security, which determines the level of complexity and antagonism of the geopolitical interests promoted by all geopolitical actors interested in a regional geopolitical space, being released the opportunities and the constraints of geopolitical nature for the development policies of the nuclear programs of the satellite states in the sphere of regional geopolitical influence.

At present, Eurasia is the continent where the geopolitical interests of all great geopolitical nuclear powers are concentrated, and dominance by a single superpower or by a coalition of two or more powers will provide the actors with the opportunity to exert the geopolitical power in order to influence the dynamics of events determining not only the continental, but also the world security, because Eurasia is a geopolitical area with most international actors who are important for the world politics, with a vast terrestrial area, a high number of inhabitants, enormous geopolitical potential, with most geopolitically contested regions and most military powers officially recognised as possessing nuclear weapons.

References

1. Huntington S (1999) The lonely superpower. Foreign Affairs 78(2):35–49. ISSN 00157120
2. Waltz K (2006) Teoria politicii internaţionale. Editura Polirom, Iaşi. 325 p. ISBN 978-973-46-0447-0
3. Juc V (2015) Securitatea naţională şi politica externă ale Republicii Moldova în contextul transformărilor structurale de sistem postrăzboi rece. F.E.-P. Tipografia Centrală, Chișinău. 200 p. ISBN 978-9975-53-530-4
4. Puşcaş V (2005) Relaţii internaţionale/transnaţionale. Sincron, Cluj. 285 p. ISBN 973-9234-58-5
5. Тузиков, Андрей Основы геополитики. Москва: КноРус, 2004. 272 с. ISBN 5-94761-033-7
6. Ikenberry J (2001) Getting hegemony right. In: The National Interest, vol 63, pp 17–24. ISSN 08849382
7. Киссинджер Г (2002) Нужна ли Америке внешняя политика? Ладомир, Москва. 352 с. ISBN 5-86218-405-8
8. Богатуров А (2003) Иракский кризис и стратегия "навязанного консенсуса". In: Международная Жизнь, vol 3, pp 30–39
9. Brzezinski Z (2000) Marea tablă de şah. Supremaţia americană şi imperativele sale geostrategice. Editura Univers Enciclopedic, Bucureşti. 238 p. ISBN 973-9436-24-2
10. Ungureanu V (2020) Redimensionarea arhitecturii geopolitice a sistemului internaţional postrăzboi rece: adversități geostrategice și replieri de securitate. F.E.-P. „Tipografia Centrală", Chișinău. 236 p. ISBN 978-9975-3430-0-8
11. Бовин А (2004) Ведущие тенденции развития международных отношений. In: Международная Жизнь, vol 4–5, pp 172–182
12. Ungureanu V (2019) Reconfigurarea arhitecturii geopolitice mondiale în contextul reedificării sistemului internaţional postrăzboi rece. In: Revista de Filosofie, Sociologie şi Ştiinţe Politice, vol 3, pp 15–36. ISSN 1857-2294
13. Ungureanu V (2019) Consolidarea dimensiunii geopolitice în noua paradigmă de asigurare a securităţii naţionale a Republicii Moldova. In: Modernizarea social-politică a Republicii Moldova în contextul extinderii procesului integraţionist european, partea a II-a / coord. Juc Victor. Chișinău: ÎS FEP „Tipografia Centrală", pp 269–289. ISBN 978-9975-3298-6-6
14. Marshall T (2018) Prizonierii geografiei. Zece hărţi care îţi spun tot ce trebuie să ştii despre politica globală. Editura Litera, Bucureşti. 288 p. ISBN 978-606-33-2508-3
15. Friedman G (2009) Următorii 100 de ani. Previziuni pentru secolul XXI. Editura Litera, Bucureşti. 213 p. ISBN 978-973-675-695-5
16. Нартов Н (2004) Геополитика. ЮНИТИ-ДАНА, Единство, Москва. 544 с. ISBN 5-238-00682-9
17. Петров В (2003) Геополитика России. Возрождение или гибель? Вече, Москва. 350 с. ISBN 5-9533-0097-2

Russian Fake News and Black Propaganda Related to Nuclear Weapons' Use in Ukraine

Marius Ilie

1 Introduction

This paper aims to identify Russian narratives in the Romanian web space that are spread through social media regarding the war in Ukraine, and by extension, the intentional or accidental use of nuclear armaments in the mentioned context. Identifying the main narratives that are used in Russian online disinformation is crucial for combating false information. Misinformation spread through social media can significantly influence public opinion and result in negative effects.

The research data was gathered from two major social media platforms. The first platform analyzed is Facebook, representing the largest social network in Romania. The second platform is TikTok, chosen due to its high number of users and the increase in registered accounts. Although Facebook has more users, user engagement with posts is higher on TikTok. As part of the research, one of the objectives is the identification of disinformation topics overlapping on both platforms will be pursued.

The main method of data gathering employed in the research entails the use of web scrapers. This method allowed the collection of a significant amount of data about users, pages, groups, and posts from the social media platforms mentioned that correlated with the subject. The methods remained consistent across both platforms, albeit with a more extensive application on Facebook. Initially, the scrapers were configured to find content based on specific hashtags on both TikTok and Facebook. The next step involved scraping multiple Facebook groups, followed by a more focused scraping operation on another set of relevant groups. Finally, a number of Facebook posts were analyzed using the direct search option.

M. Ilie (✉)
University of Bucharest, Bucharest, Romania

F.-C. Cirstoiu et al., *Medical Response Strategy in Case of Radiation Emergency Caused by the War in Ukraine*, NATO Science for Peace and Security Series B: Physics and Biophysics, https://doi.org/10.1007/978-94-024-2266-5_5

2 The Role of Social Media in Information Dissemination

In the realm of information dissemination, social networks diverge significantly from conventional media platforms like public broadcasting services (television; radio). More specifically, social networks draw in a larger audience compared to traditional media, facilitating extensive interactivity among consumers enabling mass communication on a global scale [4]. Through the development of communication platforms, the costs of reaching both national and international audiences decrease dramatically and, therefore, facilitate the spread of information. As a result, online activity, especially in the case of social networks, facilitates the creation of communities by lowering communication costs [4]. At the same time, with the ease of communication and the spread of information, social networks have become platforms that allow the rapid circulation of misinformation [4].

News and information consumption is largely influenced by the user's preferences to be selectively exposed to information that is consistent with their own political perspectives. In terms of the information that individuals are exposed to on social media, it is largely dictated by their social connections. When users browse these platforms, they are exposed to news presented with some degree of social approval, which affects the likelihood of reading such content [9]. Individuals use the perceptions, interactions and reactions of other users in their social network to determine the veracity of a piece of information.

User interaction in the comment section is an important aspect in other users' perception of the veracity of information [2]. The actions of other users in the comment section of disinformation posts significantly influence people's attitude towards disinformation, as well as their intentions to comment and share that information. Actions by other users may be more effective than warnings or other measures to counter misinformation from social media companies [2].

Thus, users have more trust in other users and less trust in the company itself. This aspect denotes several things, specifically, fake accounts (bots) can be used to comment and interact with the information to create a perception of the veracity of the information. As mentioned, warnings about the veracity of information from social networks do not have a significant positive impact in combating disinformation. Moreover, these warnings can have the opposite effect, especially for conspiratorially inclined people.

Social media networks such as Facebook and Tiktok display information that is aligned with users' preferences and beliefs through algorithms. At the same time, these networks display the material with a high degree of popularity. Fake information goes viral due to the high volume of information in relation to the low attention span of users [5]. This makes it difficult for social networks to discriminate between material containing false information and material containing correct information. Since social media is a public forum, anyone can post whatever they want without fear of being held accountable for the veracity of the information disseminated. Thus, users must decide on their own whether the material present on social networks includes false or misleading information [5].

Within social networks there is a dynamic whereby the merit of an information is associated with its popularity (the degree of interaction reported in likes, comments, shares). More precisely, there exists a direct correlation between the popularity of information on social media and the perceived accuracy of that information. Frequent exposure to an idea, often from varied sources, increases the likelihood that that information will be accepted and repeated [5]. In other words, individuals develop preferences for stimuli to which they have been frequently exposed.

The potential of information to go viral is strengthened by the negative framing of the information.

More precisely, information that is presented as negative has a higher chance of becoming viral [3]. Also, the tendency of individuals to share an economic or political news story is generally low, but the chances increase if the information is manipulated to be presented in a more negative manner. Negative news is more attractive to individuals than positive news [3]. However, it has been observed that completely false information tends to be less likely to be shared than those with some degree of truth that have been manipulated to be presented in a misleading manner. Thus, in certain cases, the effective spread of misinformation is dependent on the quality of the information [3].

Misinformation can turn into conspiracy theories. Crises, in general, because of their unpredictable nature, create optimal conditions for the proliferation of conspiracy theories. Therefore, social media has the means to create and propagate multiple conspiracy theories. Conspiracy narratives offer an alternative explanation of a given event. Often this alternative explanation is based on the categorization of an external group as omnipotent, pursuing a goal that opposes the common good [1]. Their origins can be found in people's need to deal with unfamiliar situations.

3 Methodology

The data collection process was conducted using web scrapers. Web Scraping is the process of collecting online data from social media platforms and other websites in the form of unstructured text. This technique can also be found under the names of "site scraping", "web harvesting" and "web data extraction". For the initial part in the process of data collection the web scraper was set to search posts that contained certain hashtags. The reason for using hashtags is that hashtags are tools for sorting and aggregating information by topic [6].

Visually, hashtags are characterized by a piece of text accompanied by the "#" symbol. This functionality is prevalent across most social media platforms. The function of hashtags is similar on all social networks that have this functionality. In general, hashtags are used for: marking the topic, aggregating information, socializing, expressing attitudes, initiating movements and distributing propaganda [6]. Given their characteristics, hashtags can be important tools in the dissemination and virilization of information on social media.

The keywords used in the scraping process were #zaporozhye (zaporojie); #nuclear (nuclear); #nuclearaccident (accident nuclear); #nuclearattack (atac nuclear); #nuclearwar (razboi nuclear). This scraping process was applied for both of the social media platforms.

Since Facebook has different features than TikTok, the scraping process was also applied to several public groups that were identified as disseminating false information. This decision is based on the fact that groups and pages seem to be one of the main ways in which information is shared on Facebook. Another data collection method was the direct search on Facebook of the terms "nuclear" and "russia" to further analyze potential content that was not identified using the scraper.

4 TikTok Analysis

By using a web scraper there were 1191 posts identified on TikTok [10] which had at least one of the hashtags mentioned. There is no limit for the number of hashtags that can be used on TikTok, therefore, there were posts that had up to 30 hashtags. For this reason, the number of hashtags that were analyzed was limited to only the first 10 for each post. The hashtag "#nuclear" was observed in a total of 912 instances out of the initial 1191 posts that were scraped. Notably, it predominantly appeared as the third hashtag in the order, following a format such as #hashtag1; #hashtag2; #nuclear. Its occurrence was least frequent when it appeared as the tenth hashtag. Following the data collection process [11], the pertinent posts were analyzed through manual examination.

The posts retrieved from TikTok were scrutinized in ascending order, starting from the least prevalent hashtag to the most common. Among these, the least frequent hashtag encountered was "#accidentnuclear," appearing in only 6 posts. However, half of the posts associated with this hashtag were deemed irrelevant to the research topic and consequently discarded. Moving forward, the next hashtag, "#atacnuclear," was identified 23 times. The content of posts under this hashtag predominantly focused on delineating the repercussions of a nuclear attack and suggesting potential precautions. Notably, the engagement levels for this hashtag were relatively modest. It is crucial to reiterate that this research exclusively targets content intended for a Romanian audience, hence content in other languages is not factored into the analysis.

The subsequent term under scrutiny is "nuclear war" (#razboinuclear). It is noteworthy to highlight that posts utilizing this term exhibit a substantially higher engagement rate, characterized by increased comments and likes, compared to other hashtags. This heightened level of engagement facilitates a more thorough analysis of the comments accompanying these posts. Upon interaction with the posts identified through this search, it becomes evident that the discourse within the comments encompasses both conspiratorial and religious themes.

The Zaporizhia hashtag was identified in 236 posts, being overwhelmingly found in the second position in terms of the hashtags of the posts in which it was identified.

The posts with pro-Ukraine content found in the analysis of this hashtag receive more negative interactions from users. At the same time, they have less engagement than the pro-Russian ones. Subsequent to the initial data collection on TikTok, a comprehensive analysis of the "nuclear" hashtag was not conducted. This decision was prompted by the overarching generality of the term "nuclear," which resulted in a substantial amount of data collection beyond the scope of the research.

While the term "nuclear" emerged as the most prevalent hashtag, its relevance was primarily observed when coupled with other terms. Consequently, it ceased to serve as a focal point in the analysis. The limitation of the data collection method becomes apparent in the challenge of discerning relevant content at the time of data collection through the scraper. For instance, the ubiquity of the term "nuclear" across multiple languages resulted in its inclusion in posts that fell outside the scope of the research.

5 Facebook Analysis

Facebook, like most social networks, allows users several ways to interact directly with the content they encounter. For example, the like, comment and share options have different attributes. More precisely, the like represents positive feedback to the content, the share can express the desire to increase the visibility of the information found [7]. In the case of sharing, this does not necessarily mean that the user interacted positively with the content, more precisely, a post can be shared even if the individual disagrees with the content. However, the share action increases the reach of the respective post. Regarding the comments, they reflect the opinion of a user regarding the content, they can be both negative and positive.

The data obtained from Facebook posts featuring hashtags yielded insignificant results. The majority of posts containing the nuclear hashtag proved to be irrelevant to the intended topic, highlighting the broad nature of the term. To streamline the data collection process, the search was constrained to only three of the five initial hashtags. However, even with this restriction, the method failed to yield substantial relevant data on Facebook. This observation could suggest several things. It could indicate disparities in the dissemination of information between Facebook and TikTok, or it may imply that TikTok serves as a platform where false information proliferates more readily. Consequently, a distinct approach was deemed necessary for data collection on the Facebook platform.

Facebook groups are at their simplest, content distribution channels. These are most of the time, among the most effective methods of designing information and content. There are several reasons why Facebook groups represent a favorable environment for spreading misinformation. First of all, Facebook groups usually gather individuals with similar ideas and mentalities who share beliefs, ideologies or interests. Thus, so-called "echo chambers" can be created, where misinformation can spread rapidly without being verified as members often reinforce each other's perspectives without examining the content. In essence, individuals have a tendency to gravitate towards information that aligns with their preexisting beliefs.

As previously noted, there is a significant absence of information verification, enabling rumors, misinformation, and conspiracy theories to spread. Certain Facebook groups operate as closed or private communities, effectively restricting external oversight. Within these closed groups, misinformation can circulate unchecked, devoid of external intervention or correction mechanisms.

In order to gather relevant data on the information present in groups, a scraper was used to gather posts from 6 groups in which misinformation was identified. Following the scraping process, 597 posts from the selected groups were collected. The selection of these groups stemmed from a prior observation in which posts containing disinformation, conspiracy theories, and religious content were identified. The results of the scraping were analyzed, and the posts that did not fall within the period in which the analysis was carried out were eliminated, thus resulting in 96 remaining posts. Of the 96 posts, a significant number of relevant content was not identified, therefore, 2 other facebook groups were analyzed that did not have as their main topic a religious or conspiratorial theme but rather a pro-Russian one. From this data collection, 70 posts from the 2 groups resulted.

This step was important for narrowing down the results. The engagement with the post identified was rather low compared with the engagement present on TikTok but higher than the posts identified in the previous groups. One of the initial findings from examining these posts is the favorable disposition towards actions with nuclear potential attributed to Russia. For instance, discussions pertaining to Russian nuclear armament garner more positive engagement compared to similar content concerning other countries, such as the USA. Additionally, the presence of content in the Russian language is also notable within these groups.

One interesting aspect that has emerged is that, within the analyzed groups featuring disinformation, content is predominantly generated by a limited number of users, irrespective of the group's size. Furthermore, member reactions within these groups are notably subdued, particularly relative to the overall group membership. This suggests that the engagement with false information on Facebook, as manifested through the examined public groups, is relatively insubstantial.

This finding can be explained by several factors. Either the disinformation takes place within private and closed groups, to which access is limited, or the disinformation has migrated from Facebook to other social networks. Of these two explanations, the second has a higher probability. From the first checks of the various hashtags on TikTok, a much larger number of disinformation posts were identified than on Facebook.

Lastly, using the direct search method in order to find relevant posts on Facebook yielded more consistent results. More specifically, unlike the posts that were previously scraped based on hashtags, the rate of engagement of these posts is much higher. Therefore, it could be seen once more that hashtags are not as widely used on Facebook to disseminate information effectively. At the same time, the engagement of these posts is higher than in the analyzed public groups. However, it is unclear if the engagement with the posts is higher in private groups. Several accounts have been identified that have published identical messages to each other. Comments were identified by different users publishing identical messages to other posts.

6 Comparison Between Facebook and TikTok

There appears to be a discrepancy in the number of posts and the rate of engagement between Facebook and TikTok. In addition to the discrepancy in the number of posts, there is a significant difference in the interactions, likes and comments found in the posts. As previously mentioned, the interactions in the analyzed Facebook posts are limited, especially in the case of those present in groups. Disinformation content on the Tik Tok platform is much more widespread than on Facebook.

The discrepancy between the number of results obtained from the scraping process on the two platforms is caused by the much higher volume of content available on TikTok. At the same time, this discrepancy can be further explained by the way users of the two platforms relate to hashtags. In the case of the TikTok platform, hashtags are a much more important tool in spreading content.

There are several technical factors that may explain the significant discrepancy in the volume of false information found on TikTok as opposed to Facebook. One of these factors lies in the way the TikTok platform works. This platform is based on short video content that is presented at a fast pace [8]. This quick format can encourage or influence users to consume and share content without thoroughly checking its veracity or accuracy. Also, tight time limits can negatively impact users' ability to obtain complete information. At the same time, due to the fast pace and time limitation, it is difficult to present the proper context behind a piece of information [8].

TikTok's algorithm is designed to customize content streams based on user preferences and behavior. This algorithm can lead to the creation of information bubbles or echo chambers and the amplification of disinformation content. If a user engages with disinformation material, the algorithm may continue to serve similar content to maintain user engagement and interest. Also, poor content moderation is another factor that can explain the high volume of false information present on TikTok [8].

An association was observed between disinformation propagators and their apparent level of religiosity. Many of the accounts that spread false information and conspiracies on both platforms are displaying a high degree of religiosity on their profiles. This degree is manifested in the amount of religious material or religious language used or generated. This may suggest that individuals with a high degree of religiosity are more likely to accept and spread false information or conspiracy that is consistent with their own religious beliefs. It has also been observed that religious language is often used to increase the legitimacy and emotional impact of disinformation.

In the analysis, it was also observed the use of conspiratorial language to manipulate public opinion and generate feelings of distrust. Often misinformation on social media is designed to generate intense emotional reactions and encourage division and polarization within societies. These messages can create tension and conflict and amplify existing fears and prejudices. One such religious message that seems to fit in this context is that of "the apocalypse".

7 Identified Narratives

The first narrative observed on the posts that were analyzed encourage a view that diminishes Russia's responsibility regarding the war in Ukraine, with blame often directed towards the United States or NATO. Furthermore, posts mentioning Russia's nuclear weapons tend to be viewed more positively by the audience. Concurrently, efforts are made to downplay Russia's accountability in the event of a nuclear incident. In addition to the aforementioned, conspiracy narratives mentioning the New World Order have also been identified. Within this narrative, accounts often display conspiratorial, religious, and pro-Russian tones. Another narrative suggests that the United States seeks direct conflict with Russia, viewing the war in Ukraine as a means to achieve this objective.

Among narratives with a religious tone, the most prevalent is the belief that God is on Russia's side. These narratives range from Russia defending the Christian world to framing the war as a battle against a satanic global cabal. Thus, various shades of religious narratives are combined. Another identified narrative suggests that US or NATO support towards Ukraine could trigger a nuclear war, prompting calls to reduce or halt such support. According to this narrative, Russia's use of nuclear weapons would only serve as a defense against a U.S. attack. An important aspect to note is that there were identified accounts that create and post content using multiple of analyzed hashtags, sharing common characteristics such as spreading religious and conspiratorial messages alongside pro-Russian propaganda.

The identified narratives share a resemblance in the message they convey, often seeking to absolve Russia of responsibility in triggering a nuclear incident. Regarding the potential for nuclear conflict or incidents, the message appears to evolve over time. An early narrative suggested potential defensive action by Russian forces to protect annexed territories in Ukraine, justifying the use of nuclear weapons as a defensive measure.

Another narrative implies that Ukraine spreads propaganda to incite NATO involvement in the conflict, positioning Russia as an unwilling participant and at the same time, fearing a nuclear attack orchestrated by Ukraine. Yet another narrative suggests that Ukraine attacked the dam supplying water to the nuclear plant with the help of its allies to frame Russia for the attack. This incident is argued to be perpetuated by entities controlling Ukraine from behind the scenes. Once again, conspiratorial elements are added to legitimize or enhance the emotional impact of the message. The analyzed content converges on the common theme of diminishing Russia's potential responsibility in any nuclear incident.

References

1. Buturoiu R et al (2021) Who believes in conspiracy theories about the COVID-19 pandemic in Romania? An analysis of conspiracy theories believers' profiles. Societies 11(4):138. https://doi.org/10.3390/soc11040138
2. Colliander J (2019) "This is fake news": investigating the role of conformity to other users' views when commenting on and spreading disinformation in social media. Comput Hum Behav 97:202–215. https://doi.org/10.1016/j.chb.2019.03.032
3. Corbu N et al (2020) Does fake news lead to more engaging effects on social media? Evidence from Romania. Communications 45(s1):694–717. https://doi.org/10.1515/commun-2019-0152
4. Flew T, Iosifidis P (2019) Populism, globalisation and social media. Int Commun Gaz 82(1): 7–25. https://doi.org/10.1177/1748048519880721
5. Hilary IO, Dumebi O-O (2021) Social media as a tool for misinformation and disinformation management. Linguist Cult Rev 5(S1):496–505. https://doi.org/10.21744/lingcure.v5ns1.1435
6. Laucuka A (2018) Communicative functions of hashtags. Econ Cult 15(1):56–62. https://doi.org/10.2478/jec-2018-0006
7. Quattrociocchi W, Scala A, Sunstein CR (2016) Echo chambers on facebook. SSRN Electron J. https://doi.org/10.2139/ssrn.2795110
8. Schellewald A (2021) On getting carried away by the TIKTOK algorithm. AoIR Selected Papers of Internet Research. https://doi.org/10.5210/spir.v2021i0.12039
9. Tucker J et al (2018) Social media, political polarization, and political disinformation: a review of the scientific literature. SSRN Electron J. https://doi.org/10.2139/ssrn.3144139
10. Web scraping, data extraction and automation APIFY (2023) Apify. Available at https://apify.com/

Data

11. https://docs.google.com/spreadsheets/d/1RgLObr2l8SdWi1D_k8IULiBaX9dHUQzMTuTLgTAQeQw/edit?usp=sharing

Probabilistic Foundations of Nuclear Risk Assessment

Davide Barbieri

Abstract Risk analysis is a fundamental task of intelligence analysts and engineers, as it assesses the possibility of some threat to the safety and security of a country, of its critical infrastructures and of its people. Unfortunately though, when risk analysis is performed with a qualitative approach, it is often ambiguous and vague. Therefore, a more structured and quantitative method is required in order to assess the risk in a way that it is immediately understandable and actionable for decision and policy makers. This is particularly true for nuclear risk, as it is one of the major threats to the security and safety of the national community. For these reasons, in this study a probabilistic framework for nuclear risk assessment is presented, adopting techniques and methods from both descriptive and inferential statistics.

1 Introduction

Nuclear events can have catastrophic outcomes, ranging from the destruction of critical infrastructures (like energy and power plants, for both civilian and military purposes) to the loss of human lives. In this case, casualties may result as a direct consequence of a nuclear explosion or as a later outcome of the exposure to radiations, which may provoke deaths even for years after the accident.

For these reasons, and in order to prevent the harshest effects of a nuclear event, risk analysis must be performed. Risk can be defined as the probability that something adverse will happen within a given period of time, usually in the near future, and in a given place. Risk is therefore a probability, but it takes into consideration also the damage or impact of the adverse event in case it happens. The probability that something not harmful happens is therefore not related to risk.

Risk analyses are the task of both engineers and intelligence analysts. Often, such analyses are performed in qualitative terms, which may result in ambiguous, vague

D. Barbieri (✉)
Università degli Studi di Ferrara, Ferrara, Italy
e-mail: davide.barbieri@unife.it

F.-C. Cirstoiu et al., *Medical Response Strategy in Case of Radiation Emergency Caused by the War in Ukraine*, NATO Science for Peace and Security Series B: Physics and Biophysics, https://doi.org/10.1007/978-94-024-2266-5_6

or inconsistent results and reports. In qualitative analysis, categories like low, medium or high can be associated to both probability and damage, making it difficult for decision and policy makers to act coherently according to the analysts' conclusions.

Therefore, quantitative risk assessment should be preferred, as it is already performed by different organizations of NATO countries. In particular, the US Nuclear Regulatory Commission (NRC) performs Probabilistic Risk Assessment (PRA) [1]. Nonetheless, there are several opportunities for improvement, adopting statistical techniques and methods, and implementing the proper software which is needed to assist analysts in their tasks.

The aim of this study is to describe a methodological framework which can be adopted in order to assess the occurrence of nuclear events worldwide, and in order to perform a predictive analysis of the risk of a nuclear event. As an example for the latter, the nuclear plant close to the Ukrainian city of Zaporizhzhia is taken into consideration.

2 Assumptions and Definitions

2.1 Impact

It is not straightforward to establish the potential damage in case of an adverse event. In some domains, like cybersecurity, it can be quantified in terms of monetary costs (for example in case of a ransomware attack) even if estimates can have a high degree of approximation. In the case of nuclear events, a monetary assessment can be eventually made for the loss of infrastructures, like reactors, or even civilian buildings surrounding the area. In other cases though, especially when the lives of people are at stake, defining costs and damages in case of an adverse event is more complicated and unavoidably involves discussions of ethical nature on the cost of a human life.

Often, analysts resort to a categorial variable, like a scale from low to high, to evaluate damage. A possible way to quantify the impact of nuclear events is to adopt the International Nuclear and Radiological Event Scale (INES) [2], which varies from 0 (no event) to 7 (major accident), even if is not so obvious what the rating descriptions mean, as they can be fairly subjective and approximate. Ultimately, INES gives a numeric, discrete index, which is not very different from a qualitative variable.

For the sake of this study, in the context of nuclear risk, we have considered any event potentially very harmful, even if it is not necessarily the case. As a consequence, a system, like a nuclear plant, will either behave as expected—possibly most of the time—or not. The latter case can be regarded as a potentially disastrous security or safety breach. Our problem is therefore binary: we are interested at estimating the probability that either a plant will function as expected, or that a fairly rare but potentially very harmful event will happen in the near future.

2.2 Probability

Probability can be expressed in the same qualitative terms as impact, as low, medium or high. Still, qualitative assessments are far from rigorous, as they are often ambiguous and vague. Heuer [3, p. 155] gave an interesting example: a sample of 23 NATO military officers, accustomed to reading intelligence reports, were asked to quantify numerically (using a number between 0 and 100, or a percentage) different expressions of probability, like "it is highly unlikely that…". The results showed a large diversity in their interpretations. Further, analysts can hardly be held accountable for their work, if they express their estimates qualitatively. In fact, almost everything can be considered more or less probable, making the analysts' conclusions quasi tautological and therefore not easily falsifiable. As a consequence, many organizations prefer to adopt quantitative risk analysis, based on numeric probabilities.

The classical definition of probability is: $p =$ *(number of outcomes of interest)/ (number of possible outcomes)*.

It may be considered the "ideal" or theorical probability, which only works when all possible outcomes of a trial or process are equally likely (i.e. frequent). Since we are addressing rare events, the frequency of which is not known or defined in advance, or may differ greatly, this assumption does not hold.

More often, a definition of probability based on frequency is given: $p = f(E)/n$.

where f is the frequency of an event in the past, E is the event of interest and n is the (possibly large) number of observations. Contrary to the previous, this is an empirical (i.e. based on evidence) definition of probability, which bears nonetheless some limitations. In particular, the frequency of any event may change with time, and with the time frame taken into consideration. Therefore, the fact that something happened with a certain frequency—be it low or high—in the past, does not imply that it will happen with the same frequency in the future. Further, we are not interested in estimating the absolute or simple probability that something is going to happen, but the probability that it is going to happen given the current conditions or situation [4].

We may have therefore to perform a subjective evaluation of the conditional probability of an adverse event, given a base rate (the simple probability) from which our predictions may vary considerably given the quantitative assessment of the observed facts, that is of their *likelihood*, the recorded frequency of an event in the past.

Even if in plain language likelihood and probability may be considered synonymous, they differ in statistical terms, because the former is a historical frequency, something given, while the latter is an estimate for something which may happen in the future, i.e. it is a prediction about something we do not know yet. Therefore, it must be acknowledged that past frequencies are not predictive values, even if they are related.

Furthermore, the frequency of an event in the past may be unknown, or the current conditions under which the events are observed may be too different from those of the past, so that even likelihoods must be evaluated subjectively on the basis of the current, up-to-date, knowledge of the analysts about the event of interest.

3 Methods

The data analysis process adopted in this study can be summarized as follows: (i) problem definition, (ii) data collection, (iii) data preparation and analysis; (iv) reporting. This last step should be followed by decisions and policy making, but these are not the tasks of analysts. Beside the lack of authority, it is important that those who perform the analysis are not the same as those who take decisions on the basis of the analysis' results, as—in such a case—results may be distorted or hindered by confirmation biases.

The problem at hand can defined as the estimation of the probability of a nuclear event. Data were acquired semi-automatically from open sources by means of a basic web scraping tool. In total, 133 events were recorded in 69 years from 1952 to 2020 worldwide. They were subsequently cleaned and structured into a normalized dataset, using MS Excel [5]. Basic descriptive statistics were calculated on both time series and cross-sectional data, using the historical frequencies of observed nuclear events. Predictive analysis was performed for the plant of Zaporizhzhia, using relevant intelligence collected from open sources. Initial estimates, based on a subjective and *a priori* evaluation of the risk, were coherently updated according to the analyst's assessment of the given evidence. Results were presented mostly as graphic charts.

Assessing the risk of a nuclear event consists in calculating the conditional probability that it happens given the available evidence. In order to do so, the rule of Bayes was applied: $p(H|E) = p(E|H)p(H)/p(E)$, where p(H|E) is the probability that H is true given evidence E (this probability is in fact what we want to know); p (E|H) is the likelihood of observing E in case hypothesis H is true; p(H) is the simple, *a priori*, probability of H (its initial estimate), and p(E) is the probability of observing E, regardless of the hypothesis. These calculations can be easily performed by means of a computer, once the analysts have assessed the likelihoods of the observed evidence. The time range under consideration must be agreed upon and stated explicitly at the beginning of the analysis, and it should be limited, as the probability that something happens heavily depends on it, and in the long run we can assume any event—even the most impactful—to be very probable. For the sake of this study, a time range of a week was considered.

4 Results

4.1 Descriptive Statistics

Summary descriptive statistics are shown in Table 1.

The time series of the recorded events is shown in Fig. 1. A 5-year mobile average curve was fitted to raw data. No trend was evident from this first analysis, even if the

Table 1 Descriptive statistics

Statistics	Value
Nr of events	133
Years	69
Min	0,0
Mean	1,9
Median	1,0
Max	7,0
Variance	3,5
Range	7,0

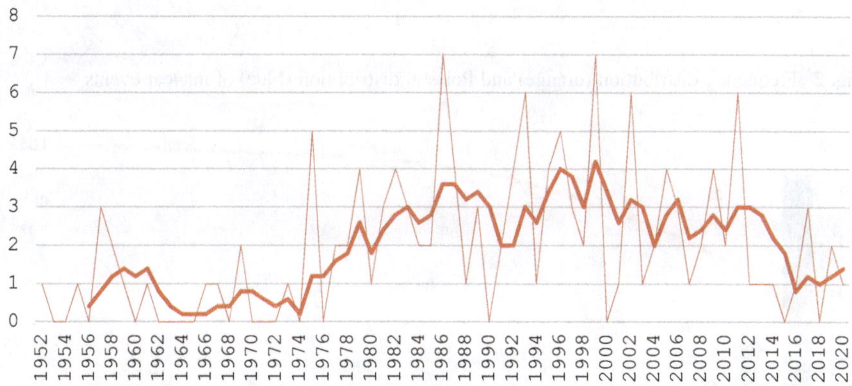

Fig. 1 Nuclear events by year (orange) and 5-year mobile average (blue)

frequency of the observed events is higher in the years between 1980 and 2010, after which they appear to diminish, even if slightly.

The frequency distribution of events is shown in Fig. 2. As it is often the case with rare events, they tend to exhibit a Poisson distribution, even if in our case they are even rarer, presenting more years than expected with no events.

Events were then analyzed as a cross-section by country. Results are shown in Fig. 3. As expected, the data present a Pareto-like distribution. The USA is the country with most recorded nuclear events in the considered period. Together with Japan, France, Canada and the United Kingdom, they present the 78.2% of the total observed events. Obviously, the overall risk depends on the number of active reactors per year, by country, but these data were not available to the analysts at the time of the study.

4.2 *Predictive Analysis*

The risk of a nuclear event at the Zaporizhzhia power plant was evaluated. Two competing hypotheses were considered: the null hypothesis H_0, that no nuclear event

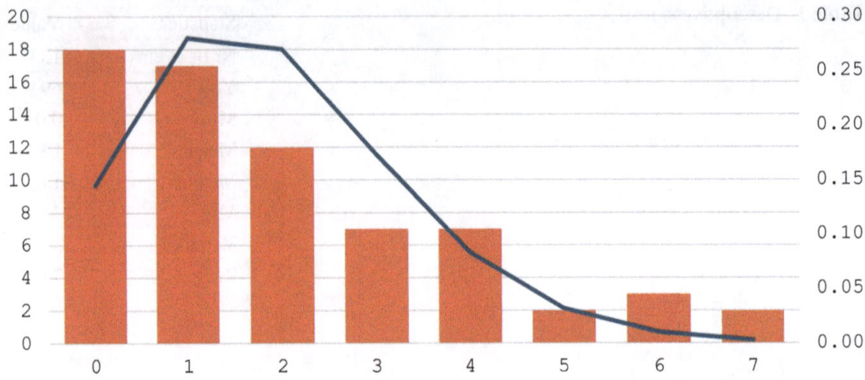

Fig. 2 Frequency distribution (orange) and Poisson distribution (blue) of nuclear events

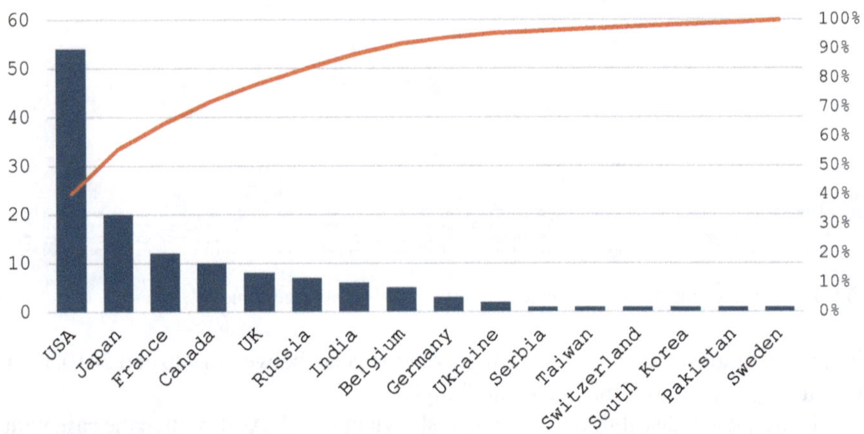

Fig. 3 Nuclear events by country (blue) and cumulative frequency (orange)

was going to happen in the week after the date of acquisition of the evidence, and the alternative hypothesis H_1 that a nuclear event would occur within a week. Incoming evidence was collected from open sources, in particular from news websites in the English language. Each piece of evidence was recorded on the date it was published (Table 2).

To start with, the analysts had to give a prior probability to the null hypothesis. The alternative hypothesis would necessarily take 1 - $P(H_0)$ as a value. The likelihoods of the evidence in case of no event $P(E|H_0)$ and in case of the event $P(E|H_1)$ were individually evaluated by the analysts, on the basis of their personal experience on the subject. Posterior conditional probability was then calculated automatically by the spreadsheet. Results were charted, for each alternative hypothesis, and presented as in Fig. 4.

Table 2 Evidence collection and likelihood evaluation

Date	$P(H_0)$	$P(H_1)$	Evidence	$P(E\|H_0)$	$P(E\|H_1)$
07/07/2023	0,80	0,20	IAEA: No mines close to plant	0,90	0,40
25/07/2023	0,90	0,10	IAEA: Mines close to the plant	0,20	0,50
03/08/2023	0,78	0,22	IAEA: Zaporizhzhia plant out of control	0,30	0,60
10/08/2023	0,64	0,36	Drone attacks to Zaporizhzhia	0,45	0,60
11/08/2023	0,57	0,43	Water spill	0,30	0,50
22/08/2023	0,45	0,55	Explosion near plant	0,35	0,45
04/09/2023	0,39	0,61	Russia lost first line of defense near Zaporizhzhia	0,70	0,30

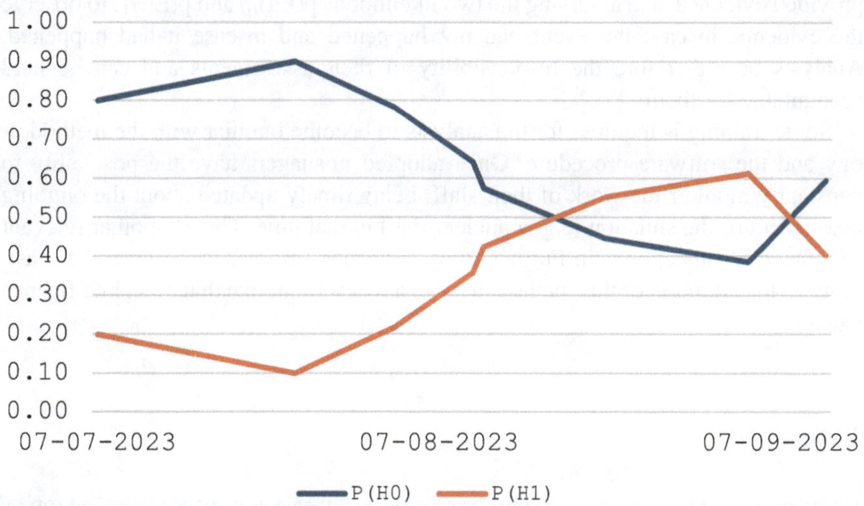

Fig. 4 Risk trend

5 Discussion and Conclusions

This study acknowledged the importance of describing in quantitative terms the results of a risk analysis. It presented a framework for risk assessment which can be used to quantify the risk of any adverse event, including the nuclear case. The descriptive part can be used by analysts to assess the *a priori* probability (the initial estimate) of the event under consideration. Subsequently, the analysts can quantify the *ex-post* risk, assessing the likelihood of the provided intelligence. This study also showed how to present the analysis and the results graphically, by means of charts which are commonly used in statistics.

In the course of the practical part, when evidence was collected and evaluated, the analysts had to resort to a subjective—i.e. non strictly mathematical—assessment of its likelihood. Even if this may appear as a limitation of the study, it simply acknowledges the fact that an objective probability, which can be calculated applying a predefined equation, is something hard to find in the empirical world. Most often, likelihoods—and therefore the subsequent posterior probabilities—cannot be calculated as relative frequencies of historical events. The main reason for this is that in some applications of risk analysis, both the event under consideration and the collected evidence, either in favor or against the event, can be rare and not observed frequently in the recent past.

The risk assessment of nuclear events can be an example of such a situation. Beside the fact that the event itself to be predicted is rare, the collected evidence is not necessarily related to other events of the same kind which happened previously. The analysts must therefore resort to their own personal experience, weighing the provided evidence, and assessing the two likelihoods $p(E|H_0)$ and $p(E|H_1)$ to observe the evidence in case the event had not happened and in case it had happened. Analysts bear therefore the responsibility of their assessments and can be held accountable for them.

Some training is required for the analysts to become familiar with the methodology and the software procedure. Once adopted, managers have the possibility to constantly monitor the work of their staff, being timely updated about the ongoing assessment of the situation (e.g. a nuclear risk) in real time. This is another relevant advantage of this approach. Further study is needed in order to evaluate the feasibility of the adoption of this methodology in a real setting, like that of an intelligence agency.

References

1. USNRC United States National Regulatory Commission, Probabilistic Risk Assessment (PRA). https://www.nrc.gov/about-nrc/regulatory/risk-informed/pra.html. Accessed 29 Nov 2023
2. IAEA International Atomic Energy Agency, International Nuclear and Radiological Event Scale. https://www.iaea.org/resources/databases/international-nuclear-and-radiological-event-scale. Accessed 29 Nov 2023
3. Heuer RJ (1999) Psychology of intelligence analysis. Center for the Study of Intelligence
4. Barbieri D (2017) How to excel at intelligence analysis, XXIII international conference on intelligence in the knowledge society, Bucharest, Romania, 26th–27th October 2017
5. Barbieri D (2013) Bayesian intelligence analysis, in proceedings of the 19th conference on intelligence in the knowledge society, Bucharest, Romania, 18th October 2013

Russian Provoked Nuclear Disaster, Attack or Accident: Scenarios, Probability and Consequences

Iulian Chifu

Abstract Nobody in good faith can rule out the possibility that a nuclear weapon will be used in Ukraine or the possibility of another type of nuclear disaster. But the probability of such a scenario is quite low, as the impact to the war is meaningful, except if there is a (improbable) side effect of such a use in the resolutions and perception about the fight of the Ukrainian public. What we have learnt up to now is that the societal resilience as well as the intangibles of war are showing that Ukraine will not give up. In this situation, accidents can happen, but an event of such kind involving nuclear material could be a rare one with very low probability. The current article explores both nuclear attack, attack at a nuclear power plant or nuclear waste deposit, the probability, and the possible consequences in each case.

1 Introduction, Methodology, Approach

We are approaching the potential nuclear impact on Russia's war of aggression in Ukraine based on two possible events: an assumed use of a nuclear bomb on the front line or an accident at a nuclear facility in Ukraine that could result in a nuclear leak of radiological material of a certain importance. In the second case, we are using an impact study based on AEIE criteria for nuclear safety and its reports on the ground in Ukraine, as well as technical based studies in such cases. For the use of a nuclear bomb, we made an extensive and specific analysis of Russia's documents and previous behaviour in nuclear crisis related to Russia's escalation philosophy and strategy and the escalate to de-escalate concept, an analysis of the Russian deterrence concept. Here we are looking at the logics of deterrence and coercion through nuclear threat continued role in wartime, as well as the meaning of the nuclear threat saber rattling by a number of high Russian officials. We add to this the studies on Ukraine's resilience in front of a large-scale high-intensity long-term war proven by

I. Chifu (✉)
"Carol I" National Defence University, Bucharest, Romania
e-mail: Iulian.Chifu@snspa.ro

F.-C. Cirstoiu et al., *Medical Response Strategy in Case of Radiation Emergency Caused by the War in Ukraine*, NATO Science for Peace and Security Series B: Physics and Biophysics, https://doi.org/10.1007/978-94-024-2266-5_7

societal security characteristics and the intangibles of war, identified for the Ukrainian society in the first 2 years of war.

The use of a nuclear bomb in Ukraine has been assessed as a game changer. But not the way that the decision makers that would approve this use would expect. It would represent the most significant game changer in the state of international relations since the Second World War [1], because Russia's use of nuclear weapons is an implicit admission of an eventual conventional defeat in Ukraine, but it will represent also a complete exit of Russia from the world system, with major costs of real isolation from the international community. We would no longer be talking about the autocracy versus democracy rivalry, but about a maverick actor in the system that challenged the most important elements and agreements about the balance of power and efforts to avoid war. Russia would have difficulties to explain such an escalation to its partners in the so-call Global South, but first to China, one of the partners that explicitly warned against such an use. Teaming up with Russia in any direction would be morally impossible and would enshrine huge costs in terms of legitimacy for the country that would align with it after the use of a nuclear weapon in war.

We cannot rule out that, moreover, since we are talking about a permanent member of the Security Council that has completely escaped all logic and control, other international costs would come. China, India and the states of the Global South would find it very difficult to support Russia in any way, and its expulsion or suspension of the veto rights in the Security Council and the containment into a complete isolation would be matched only by the effort to support and assist Ukraine to liquidate the consequences of such an event. The allies and partners as well as the former non-aligned countries, but even China would have to have a participation in this effort. In terms of the war, Ukraine would most probable receive all the support it would need, including the presence of NATO troops directly on the ground, to first liquidate the consequences of such a strike and to obtain the liberation of its territory.

2 The War and Nuclear Facility Accident or Attack: A Low Probability Incident

An incident at a nuclear facility would have an impact around the site of the place where such an event would occur. Ukraine has four nuclear power-plants: Khmelnytskyy, Rivne, South Ukraine and Zaporizhya, in addition to the Chernobyl NPP, which has both a wet and a dry spent fuel storage facility. The Kharkov Institute of Physics and Technology (KIPT) has a Neutron Source installation in Kharkiv and the research reactor at the Institute for Nuclear Research in Kyiv use and store nuclear material [2]. The only nuclear power plant that was in the line of fire is the Zaporizhya Nuclear Power Plant, the biggest in Europe, moved under Russian control. Despite these changes, the Ukrainian nuclear staff continue to operate the ZNPP. United Nations General Assembly adopted the resolution A/RES/ES-11/4 which states that this power plant is considered an Ukrainian one [3].

An accidental conventional impact at a nuclear power plant hitting one of the reactors, is practically without any direct consequences since the level of security prevents the possibility of breaking inside the corps of the reactor. Only a planned, high intensity, multiple missile shots targeted attack in a relatively long period of time could harm the building protections and the reactor ensemble to get inside the core and let radiation spread around. The probability of such an event is extremely low and would enshrine direct responsibility of the party that does such an attack.

The IAEA works based on seven pillars of security, safety and safeguards related to nuclear facilities: physical integrity—reactors, fuel ponds or radioactive waste stores must be maintained; safety and security systems and equipment—fully functional at all times; operating staff able to fulfill their safety and security duties and have the capacity to make decisions free of undue pressure; off-site power supply for each nuclear power plant; logistical supply chain—uninterrupted, with transportation in and out; radiation monitoring and emergency preparedness and response, effective, on site and off site, communications—with regulator and others, at all times [4].

In the case of Ukraine, the IAEA has identified major impacts and challenges concerning all seven pillars. And that in both Chernobyl and especially Zaporizhya nuclear power plant. The last report [5] as well as the consecutive assessments and updates on the matter [6] are introducing important warnings, concerns, and noted numerous breaches of the international regulation where Russia is a part, but not a sense of danger and perspective of an accident with catastrophic consequences. Even the fact that the monitoring is very strict, with officers of the IAEA on the ground and numerous reports, as well as the cooperation that continued with the Russian authorities could be a prove for a very low probability of such accidents in this case, with local and minor impact expected.

On another point, hitting with a conventional weapon a nuclear waste storage has a different consequence. It has more or less the same high protection as that of a nuclear power plant, after the events in Chernobyl, but we have to consider two different stages of evolution, according to the specialists questioned by this author: if it is older than 7 years, there are no negative consequences, being equivalent to a dirty bomb *in situ*, with consequences around the area of the potential explosion and relative of the dust taken by the winds, on smaller distances. If there are less old than 7 years, the impact is already more important, with the possibility of the radiation of being spread on longer distances and have a bigger effect on areas situated at up to several hundred kilometres away [7]. But a Fukushima or Chernobyl type of event is one with low probability to be repeated in Ukraine.

3 The Use of Nuclear Weapons in War: Deterrence, Defence, Retaliation

The use of nuclear weapons is a part of any military doctrine of countries that have them. It isn't worth having such a formidable weapon and not having the rules of engagement in place or a doctrine how to use them. On moral basis, nuclear weapons should be a deterrence tool and not designed to be used. But for the same deterrence reasons, a public transparent doctrine describing in what conditions and how the nuclear capabilities are used is a must. And here, again, moral basis is creating an environment where those types of extraordinary weapons should be used only in defense. But countries, for the same reasons of credible deterrence, have embraced rules to use in offensive, or aim at least to have ambiguity related to this use. Another clear delimitation is that between the use of strategic weapons and non-strategic nuclear weapons.

In the case of Russia, all the public statements of military and political officials before the current war—and therefore not subject to the use of nuclear saber rattling for informational warfare matters—are underlying the defensive character of the nuclear arsenal. Valeriy Gerasimov, Chief of the General Staff, has stated that Russia was obliged to "create a threat in response to threats" [8], in relation to the nuclear weapons of the US. Vladimir Putin said in 2016 that "nuclear weapons are a deterrent and a factor of ensuring peace and security worldwide. They should not be considered as a factor in any potential aggression, because it is impossible, and it would probably mean the end of our civilization" [9].

Russia's formal doctrine considers nuclear weapons designed primarily as a deterrent or in a defensive context. So, the first use of strategic nuclear forces is not present in the documents, but nuclear capabilities are intended for a second strike that "would inflict unacceptable damage on an aggressor" [10]. They are designed as a retaliatory strike, but only when there is a clear certainty that a strategic nuclear attack has been proved by the early warning systems. In October 2018, the Russian Strategic posture included the explanation of Vladimir Putin that "only when we know for certain—and this takes a few seconds to understand—that Russia is being attacked, we will deliver a counter strike. This would be a reciprocal counter strike. Why do I say "counter"? Because we will counter missiles flying towards us by sending a missile in the direction of an aggressor. Of course, this amounts to a global catastrophe but I would like to repeat that we cannot be the initiators of such a catastrophe because we have no provision for a pre-emptive strike" [11].

For sure, all those are declarations, statements, and public official positions. In that field, it is very difficult to play with words. But in a world where we recorded a fall of arms control in Europe, that perspective could not be enough for anybody. Everything we knew from the Cold War is lost: the United States withdrew from the Anti-Ballistic Missile (ABM) Treaty in 2002, Russia retreated from the Conventional Armed Forces in Europe Treaty in 2006, effective 2007 and stopped respecting its provisions in 2015. The Intermediate Nuclear Forces Treaty ended in 2019 due to the retreat of the United States after evidence and repeated discussions about breaches made by the Russian Federation, the United States withdrew from the Open Skies Treaty in 2020 due to blockages in applying it on Russian occupied territories in Europe, and therefore Russia withdrew itself in 2021. And in 2023, Russia suspended cooperation with the United States on a New Strategic Arms Reduction Treaty (START) [12].

4 The Informational Warfare and the Statements Related to the Use of Nuclear Weapons

We have seen lately an enormous quantity of statements, at different levels, about the use of nuclear weapons. If the vice-president of the Security Council, Dmitri Medvedev, had assume a role in that area, with some dozen statements and threats to use nuclear weapons, as a self-assumed role, in his speech for the State of the Nation on February 21st, 2023, Vladimir Putin himself alluded to an indirect threat to the use of nuclear weapons that he linked directly to the war in Ukraine:

> As if there is no connection between strategic offensive weapons and, say, the conflict in Ukraine (...) them seeking to inflict a strategic defeat on us. This is either the height of hypocrisy and cynicism, or the height of stupidity, but they are not idiots. They are not stupid after all. They want to inflict a strategic defeat on us and also to get to our nuclear sites [13]

This statement received a harsh reaction from the White House. President Joe Biden answers the claim of an existential threat:

> The United States and the nations of Europe do not seek to control or destroy Russia. The West was not plotting to attack Russia, as Putin said today. And millions of Russian citizens who only want to live in peace with their neighbors are not the enemy. This war was never a necessity; it's a tragedy. President Putin chose this war. Every day the war continues is his choice. He could end the war with a word. It's simple. If Russia stopped invading Ukraine, it would end the war. If Ukraine stopped defending itself against Russia, it would be the end of Ukraine. That's why, together, we're making sure Ukraine can defend itself [14].

One of the most important explanations about this nuclear rhetoric is linked to informational warfare. The idea is to establish a belief among many Western observers that Russia can and will use nuclear weapons so that Russian threats for the use of nuclear weapons become credible. Therefore, the very concept of strategic deterrence is linked to the concept of reflexive control, a Russian strategy to determine the reactions of the opponent based on an induce stimulus.

This scope of the concept clearly includes the entire domain of information warfare [15]. It is linked to an intimidation culture characteristic to Putin's Russia. So those statements are vivid parts of the strategic deterrence in action and are superposed with the promotion of a Western deterrence that is not seen as credible: therefore, the balance of deterrence has been destabilized [16]. Moreover, since the concept of strategic deterrence assumes the existence of a condition of near war, if not actual war, at all times, in its hybrid, informational or even concrete military form by extending the war of aggression in Ukraine as being a war against NATO and the "Collective West".

So, the official Russian definition of the term of strategic deterrence (strategicheskoye sderzhivanie) is: "A coordinated system of military and non-military (political, diplomatic, legal, economic, ideological, scientific-technical and others) measures taken consecutively or simultaneously with the goal of deterring military action entailing damage of a strategic character. Strategic deterrence is directed at the stabilisation of the military political situation in order to influence an adversary within a predetermined framework, or for the de-escalation of military conflict. The objects to be influenced through strategic deterrence may be the military-political leadership and the population of the potential adversary state (or coalition of states). Strategic-deterrent measures are carried out continuously, both in peacetime and in wartime [17]. But the whole system does not stop only at the level of some first-generation informational warfare but is moving to a more solid use. "Russian military planners [have been] pursuing a broad range of upgraded and new versions of nuclear weapons," which "suggests that the[ir] real doctrine goes beyond basic deterrence and toward regional warfighting strategies, or even weapons aimed at causing terror" [18].

According to Stephen Blank, strategic deterrence definition aims at imposing Moscow's control on the entire process of deterrence, encompassing pre-war and wartime periods, if not postwar times as well [15], it does not subsume only the cross-domain coercion, but it also presupposes constant tensions that could explode at any time into actual hostilities [19]. The third unique feature of the Russian deterrence concept is its blending of the logics of deterrence and coercion through its continued role in wartime, erasing distinctions between offense and defense and between compellence and deterrence [19]. The Russian concept seeks more actively to influence wartime calculations through demonstrating Russian willingness to use coercive measures. The need to demonstrate coercive capability increases as the destructiveness of the deterrent tools employed decreases [17].

5 The Conditions of First Use of Nuclear Weapons

Alexey Arbatov underlined that the nuclear balance is so stable and for some time that limited or selective nuclear strikes are again considered possible, and not necessarily leading to a massive exchange. It is call "tailored nuclear options for limited use" [20]. Or limited strategic nuclear strikes. That would be the current deterrence paradox and the folly of limited nuclear war. Russia's current Military Doctrine notes that "the most important task of the military is guaranteeing the air-space defence of key sites in the Russian Federation and [ensuring] readiness to repel an air-space attack" [21]. At the same time, a selective use of strategic nuclear systems is envisioned at a certain phase of the armed conflict [22].

The motives for the use of nuclear weapons are spread between showing resolve, de-escalating conflicts—a Russian form for the doctrine escalate to de-escalate, especially in a regional conflict that harms Russia's fundamental interests [21] or local conventional conflict between Russia and NATO. But in spite of the fact that nuclear weapons are considered as guarantors of peace and security among great powers—so not in regional conventional wars—other reasons for the use of nuclear weapons are:

- Russia's own great-power status that needs to be reaffirmed and proved,
- the perception that the West, the US especially, appears to reject mutual vulnerability as a stabilizing factor, and does not want to deal with Russia as an equal, but it continues to improved missile defense capabilities, but not ultimately [10].

– because Russian public opinion surveys suggest that the Russians believe that their country's use of military force in Ukraine and Syria and its assertiveness abroad have increased Russia's importance in the world [23].

In the case of Russian's approach to the perspective of the concept *escalate to de-escalate*, beginning with the Military Doctrine of 2000, Russia's official position on nuclear weapons has been that they are possible means of de-escalation [24]. This document stated that limited and targeted use of nuclear weapons would inflict "tailored damage" on the enemy and end the destructive attack—hence "de-escalation". But we cannot rule out that Russia believes that nuclear weaponry is an essential factor of state power and, since it is one of its last arguments to claim superpower status [25], it relies on it and assess its credibility by enforcing conditions of use or, at least, a designed level of ambiguity about that perspective. Modernisation of weapons is creating a level of responsibility in the nuclear field. A thorough study in that direction has been achieved with recommendations [26], but lately Russia didn't seem to rely on those same rules, prompting a reaction that could accentuate instability and risk [27].

For sure, the non-strategic use of nuclear weapons in regional deterrence is kept under a deliberate ambiguity in Russia. Russia might use nuclear weapons in the event of "large-scale aggression by conventional weapons, in situations deemed critical to the national security of the Russian Federation" [28]. Russia's Security Council Secretary Nikolai Patrushev attempted to introduce a different stance, saying that Russia would not rule out pre-emptive nuclear strikes, including in local conflict [29]. But the final text of the 2010 doctrine raised the bar for nuclear use. This may suggest that those arguments failed to hold sway with senior civilian leaders [30]. The 2014 doctrine added references to non-nuclear deterrence but didn't accentuate the first use [31].

The naval doctrine insists not on escalate to de-escalate, but on "demonstration of readiness and determination", so the main purpose of the Russian non-strategic nuclear weapons could be more to provide the Russian political leadership with a range of flexible alternative options of action and to help them maintain an environment of ambiguity supporting Russian deterrence [32]. But the Russian military-analytical community is divided in what their non-strategic nuclear weapons should be for [10].

On another point, some put the maintenance of the use of Russian non-strategic weapons on advantages of the Western countries on precision guided weapons. If conventional weapons are insufficient, Russia could employ non-strategic nuclear weapons, considering that the use of very- low- and ultra- low- yield nuclear weapons will not trigger a strategic nuclear exchange. So, there is a place for a limited strategic nuclear response [33]. Russian research and development into ultra-low- yield (ULY) and very- low- yield (VLY) nuclear warheads in the range of tens to hundreds of tons of high- explosive equivalent for non-strategic nuclear weapons (NSNW) reduces the barrier to use [34].

Central Intelligence Agency analysis of Russian thinking on ultra- low-yield and very- low- yield nuclear weapons from August 2000 suggests "the need for subkiloton nuclear weapons with minimal long- term contamination had been argued in the media by senior Ministry of Atomic Energy (Minatom) officials, nuclear weapons scientists, and military academics since the mid-1990s". The same report noted "very- low- yield weapons as an 'asymmetric response' to US superiority in conventional weapons" [35]. According to the Defense Intelligence Agency, as of May 2019, "Russia's stockpile of non-strategic nuclear weapons, already large and diverse, [was] being modernized with an eye towards greater accuracy, longer ranges, and lower yields to suit their potential war- fighting role" [36].

But Russia's Escalation Philosophy and Strategy assessed that the conflict between Russia and NATO is likely to reside on regional and large-scale conventional conflict, depending on the degree to which NATO strikes Russia and Russia strikes NATO [37]. The real mistake could be in the belief that there is an escalatory gap between the use of non-strategic nuclear weapons and the circumstances that create a strategic nuclear exchange [38]. So the limited employment of non-strategic nuclear weapons (NSNW) is becoming an increasingly rational choice for Russian President Vladimir Putin in Russia's ongoing war against Ukraine [39]. Russia could do that since there is a belief that Putin will fall from power if he loses the war, then that Russia will lose its great power status due to the military, economic, and reputational degradation resulting from the war, but also because the Western answer to the use of such weapon has not being clarify. Then, a Western conventional attack against missile warning radars and satellite control sites would "blind" the leadership of Russia, so it will also entail a high probability of nuclear retaliation in accordance with Russia's Military Doctrine [40].

For really making the point, we also need to also investigate the consequences and impact of such a use in the first line, in the war. Some analysts have argued that open-source analysis of exercises with dual-capable systems offers only ambiguous evidence on preparations to use a sub-strategic nuclear weapon, because they could be performing in conventional or nuclear roles [41]. Although unlikely to turn the tide of battle or achieve the political objective of conquering Ukraine, limited NSNW use could still prove Russian mettle and allow it to maintain its place on the world stage [39]. But our conclusion remains that the costs of such an action are so important that such an attack could be considered improbable. We are still maintaining the Putin factor of unpredictability, so such a use could definitely not be ruled out.

6 Ukraine's Societal Resilience and Intangibles of War as Means of Deterrence for the Use of Nuclear Weapons

Ukraine has proved its resilience and capacity to absorb the pressure even during war times and to defend its identity, existence as well as territory and population. It came with a number of tools linked to societal security [42] and especially to the intangibles of war [43]. How could this affect Russia's decision to ever use a nuclear weapon in the war? Actually, our thesis is that societal resilience [44, 45, 46, 47, 48] of Ukraine is a very clear indicator that even if there would ever be the use of a nuclear sub-strategic nuclear weapon on the ground, it wouldn't have any consequence on the will to fight of the Ukrainians and will not determine them rising the white flag. On the contrary, the process would be the opposite, combined with the side effects and international condemnation of such use.

Societal resilience, national resilience, community resilience, in general, is seen as a characteristic of the human community itself that involves both reacting to shocks and crises, and surviving, but also rebuilding society, a feature that measures the capacity to work together in such times [49].

Community resilience, in general, is defined as a property of dynamic social-ecological systems and assesses the extent to which communities can respond positively to several risks, including shocks, extreme events or other changes [49]. The characteristics identified vary from author to author [50]. They often include references to attachment to community place, leadership, community cohesion and effectiveness, established and functioning community networks, knowledge, and learning. These characteristics would be useful for community conservation, disaster risk reduction, climate change adaptation and community development [49]. Societal resilience already refers to the whole societies, it has the basic characteristics of community resilience, but raised to the level of complexity, cohesion, and sophistication of society, including the responsibilities to its members [51].

The sources of resilience are coming in the realm of societal security and intangibles since those elements explain why measurable data and quantitative models does not suffice to cover the relevance of a country in war times [52]. In terms of intangibles [43] and elements of societal security, resilience is also the probability of ensuring the task, objectives, and services during and after a crisis, civil emergency, or disruptive event. Immaterial factors, non-measurable evolutions, and events as well as the intangibles [53] could enforce and multiply the level of resilience. And these are qualitative indicators, non-measurable.

As an example, how could you measure the capacity of sacrifice? The capacity of supporting a war effort and to absorb the costs of a long-term war? The acceptability of the spendings and costs in war times, the power to endure sufferance and pain for losing loved ones in a war? Intangible war characteristics are defining, as we have seen in Russia's war of aggression against Ukraine, the one that gets a victory in a concrete clash [54], despite the measurable indicators that show a different perspective. They include the will to fight, determination, resolution, capacity of sacrifice, survival in harsh environments, for sure reputation, credibility, brand, leadership, compassion, moral, cohesion, unity, level of preparedness and experience, consciousness and support of the public opinion [55]. And some more.

Explaining the achievements on the ground during war times, in the full scale, high intensity long term war of aggression launched by Russia in Ukraine [43] lead us to studying the conundrum formed by resilience, societal security and the intangibles of war. The layered resilience that NATO has embraced [56] as concept is not enough. The difference between success and failure in the Russian war of aggression in Ukraine came from the intangibles of the war [53]. We proved the relevance and importance of this conundrum societal resilience-societal security—intangibles in war times [43].

Some of the elements revealed above are direct and tangible but the majority are especially intangible in war. The most important intangible under discussion directly related to societal security is the capacity for sacrifice—the ability to sustain a war effort and to absorb the costs of a long war, the acceptability of the expenses during war times, and the power to endure the suffering, hardship, and pain of loss of loved ones in a war. These elements are difficult or impossible to measure, to place among the tangibles of war, remaining in the space of intangibles.

The intangibles of war, those elements that made the difference between the Russian army—self-proclaimed to be the second largest army in the world—and the Ukrainian army, smaller in number but much more determined and capable of fighting for a purpose, the real Ukrainian war of independence. These intangibles relate to reputation, credibility, even the country's brand. Intangible elements, such as culture and perception, effectively shape the interpretation and evaluation of the physical realm by the cognitive realm. And the list of crucial intangibles also includes leadership, morale, cohesion, unity, level of training and experience, situational awareness and support from public opinion. And many, many others, such as the ones revealed before, including the will to fight or the determination to defend their country [55].

A very important intangible relates to legitimacy. It has multiple resorts, but the most important ones come from the democratic resilience which are influencing the need for support the defense of such desirable values and high stance in front of an aggression, itself creating deficits of legitimacy and costs in terms of ethics in international relations. It is about democratic resistance or democratic fight, the alliances and the cohesion of countries respecting the same values, principles, situated on high moral ground.

Another important point in the intangibles, combining positions and perceptions, as well, is to be on the just part, legally and legitimately justifiable side of a war, with a moral advantage, in any case having all the ingredients and components that define both a "right war" and a good side of the conflict. This deserves support and the support can come both by political and symbolic support and by direct financial or military support in a war, creating a "just cause" from supporting one side in such a conflict. A positive light will accumulate a greater sense of legitimacy and following.

Looking at the war of aggression in Ukraine, we have seen all those characteristics. So, the conclusion is not only that intangibles and elements of societal security as part of societal resilience are making the difference and are underlying the resolution of the Ukrainians, hard to be shaken even by using of a nuclear weapon. But moreover, that this is a fact already learnt by the Russian part, which creates a strong signal hard to be denied, that comes as an argument that the non-strategic use of a nuclear weapon is completely useless in this war. Therefore our final conclusion is that nuclear safety is rather granted during the war in Ukraine, with low probability of a nuclear accident at a wet storage of nuclear waste younger than 7 years and with the "Putin factor" for a low probability and no advantage low possibility use of a nuclear theater weapon in Ukraine, who could only have one purpose, showing its own public that Russia is strong and reaffirming the last element of superpower in its arsenal.

References

1. Chifu I (2023d) The impact of Russia's aggression on reshaping tomorrow's world. Buletinul UNAp, nr. 2/2023. https://revista.unap.ro/index.php/bulletin/article/view/1725/1676
2. Energoatom National Nuclear Power Company State Enterprise. https://energoatom.com.ua/en/
3. UNGA, Resolution adopted by the General Assembly on 12 October 2022, Territorial integrity of Ukraine: defending the principles of the Charter of the United Nations. https://documents.un.org/doc/undoc/gen/n22/630/66/pdf/n2263066.pdf?token=XJurIJyBi8i8aWRJOe&fe=true
4. Non Proliferation Treaty, 2020 Review Conference of the Parties to the Treaty on the Non-Proliferation of Nuclear Weapons, 4 November 2022. https://treaties.unoda.org/t/npt; https://disarmament.unoda.org/wmd/nuclear/npt/; Ali Alkis, The indispensable pillars of nuclear security and safety, 27 September 2022, at http://odcnp.com.ua/images/pdf/Pillars_Nuclear_Safety_Security.pdf
5. IAEA, Nuclear safety, security and safe-guard in Ukraine, February 2022–February 2023. https://www.iaea.org/sites/default/files/23/02/nuclear-safety-security-and-safeguards-in-ukraine-feb-2023.pdf
6. IAEA, Update 215—IAEA Director General Statement on Situation in Ukraine, no. 22/2024, 8 March 2024. https://www.iaea.org/newscenter/pressreleases/update-215-iaea-director-general-statement-on-situation-in-ukraine
7. Ministerul Energiei, Agenția nucleară și pentru deșeuri radioactive (Energy Ministry, Nuclear and radioactive waste Agency, Romania), Strategia națională pe termen mediu și lung privind gestionarea în siguranță a combustibilului nuclear uzat și a deșeurilor radioactive (National mid and long term strategy for safely managing fuel ponds and nuclear waste, august 2021. https://energie.gov.ro/wp-content/uploads/2021/08/Strategia-nationala-pe-termen-mediu-si-lung-privind-gestionarea-in-siguranta-a-combustibilului-nuclear-uzat-si-a-deseurilor-radioactive_FINALA_01.07.2021-1.pdf
8. Gerasimov V (2019) Vektory Razvitiia Voennoi Strategii [Vectors of the Development of Military Strategy]. Speech to the Academy of Military Sciences. http://redstar.ru/vektory-razvitiya-voennoj-strategii/
9. Valdai(2016)Transcript of the meeting of the Valdai international discussion Club, October 27, 2016. http://en.kremlin.ru/events/president/news/53151
10. Loukianova FA, Oliker O (2020) Russia's nuclear weapons in a multipolar world. Daedalus 149(2) Meeting the Challenges of a New Nuclear Age (Spring 2020), Published by the MIT Press on behalf of American Academy of Arts & Sciences. www.jstor.org/stable/10.2307/48591311
11. Valdai(2018) Transcript of the meeting of the Valdai international discussion Club, Sochi, Russia, October 18, 2018. http://en.kremlin.ru/events/president/news/58848
12. van Hooft P (2023) On arms control Europe should get off the sidelines. Atlantisch Perspectief 47(2) Published by: Stichting Atlantische Commissie. www.jstor.org/stable/10.2307/48732310
13. Vladimir Putin V, Presidential Address to Federal Assembly, 21 February 2023. http://en.kremlin.ru/events/president/news/70565
14. The White House, Remarks by President Biden Ahead of the One-Year Anniversary of Russia's Brutal and Unprovoked Invasion of Ukraine, 21 February 2023. https://www.whitehouse.gov/briefing-room/speeches-remarks/2023/02/21/remarks-by-president-biden-ahead-of-the-one-year-anniversary-of-russias-brutal-and-unprovoked-invasion-of-ukraine/#:~:text=So%2C%20tonight%2C%20I%20speak%20once,neighbors%20are%20not%20the%20enemy

15. Stephen Blank S (2022) Nuclear weapons in Russia's war against Ukraine. Naval War Colleg Rev 75(4) https://www.jstor.org/stable/48731998
16. Ignatius D (2022) The hidden risks of Russia's war in Ukraine. Washington Post, p A21. www.washingtonpost.com/
17. Military-Encyclopedic Dictionary of the Russian Ministry of Defense. Available at encyclopedia.mil.ru/, as cited by Kristin Ven Bruusgaard, "Russian Strategic Deterrence," Survival 58, no. 4 (2016)
18. Kristensen H, Korda M (2022) Russian nuclear weapons. Bull At Sci 78(2)
19. Adamsky D (2015) Cross-Domain Coercion: The current Russian art of strategy, Proliferation paper 54. Institut français des relations internationales, Paris
20. Arbatov A (2017) Beyond the nuclear threshold: Russia, NATO, and nuclear first use. European Leadership Network. www.jstor.com/stable/resrep17426
21. Voennaya doktrina Rossiiskoi Federatsii [Military doctrine of The Russian Federation], President of Russia, 2014. http://news.kremlin.ru/media/events/files/41d527556bec8deb3530.pdf
22. Sukhanov S (2010) Air-space defense is a task, Not a system. Air-Space Defense 29. http://www.vko.ru/koncepcii/vko-eto-zadacha-ne-sistema
23. Smeltz D, Wojtowicz L, Goncharov S (2018) Despite last year's expectations, publics sense strains in U.S.-Russia relations. Chicago Council on Global Affairs. https://www.thechicagocouncil.org/publication/despite-last-years-expectations-publics-sense-strains-us-russia-relations
24. Sokov NN (2014) Why Russia calls a limited nuclear strike 'De-escalation'. Bull At Sci. http://www.thebulletin.org/why-russia-calls-limited-nuclear-strike-de-escalation
25. Kamp KH (2015) Nuclear implications of the Russian-Ukrainian conflict. NATO Defense College. www.jstor.com/stable/resrep10312
26. Morgan A, Williams H, Responsibility N (2018) A new framework to assess U.S. and Russian behaviour. European Leadership Network. www.jstor.org/stable/resrep22126
27. Minho K. Intelligence chairman defends his hints about a Russian space weapon, The New York Times, 18 February 2024. https://www.nytimes.com/2024/02/18/us/politics/mike-turner-russia-nuclear-space.html
28. Translation of the text of the 2000 doctrine at Arms Control Association, Russia's Military Doctrine. https://www.armscontrol.org/act/2000_05/dc3ma00
29. Russia reserves pre-Emptive nuclear strike right. Reuters, October 13, 2009. https://www.reuters.com/article/idUSLD72025
30. Andrey Afanesavich Kokoshin (2014) Strategic nuclear and nonnuclear deterrence. Sci Soc 84. https://link.springer.com/article/10.1134/S101933161402004X
31. Voennaia Doktrina Rossiiskoi Federatsii [Military Doctrine of the Russian Federation], 2014 at https://rusemb.org.uk/press/2029
32. U.S. Naval War College, 2017, A 2017 Doctrinal document from the Russian navy, fundamentals of the state policy of The Russian Federation in the field of naval operations for the period until 2030, fundamentals of the state policy of The Russian Federation in the field of naval operations for the period until 2030, trans. Anna Davis (Newport, R.I.: Russia Maritime Studies Institute, U.S. Naval War College, 2017). https://dnnlgwick.blob.core.windows.net/portals/0/NWCDepartments/Russia%20Maritime%20Studies%20Institute/RMSI_RusNavyFundamentalsENG_FINAL%20(1).pdf?sr=b&si=DNNFileManagerPolicy&sig=fjFDEgWhpd1ING%2FnmGQXqaH5%2FDEujDU76EnksAB%2B1A0%3D
33. Kofman M et al (2021) Russian military strategy: core tenets and operational concepts. Center for Naval Analysis, Arlington, p 32. https://www.cna.org/
34. Boone R, Rehbein D, Swegle JA, Christopher Yeaw C (2023) Russia's nonstrategic nuclear weapons and Western air supremacy. Æther 2(1). https://www.jstor.org/stable/48714685
35. Central Intelligence Agency, Office of Transnational Issues, "Evidence of Russian Development of New Subkiloton Nuclear Warheads," Intelligence memorandum, August 30, 2000 (redacted for release)

36. Ashley Robert P Jr (2019) Russian and Chinese nuclear modernization trends. Hudson Institute, Washington, DC. https://www.dia.mil/
37. Mission and Objectives of the Russian Armed Forces. Ministry of Defence of the Russian Federation (website). https://eng.mil.ru/
38. Yost DS (2001) Russia's non- strategic nuclear forces. Int Aff 77(3)
39. Colas Brandon A (2023) *Rational choice?* Æther 2(2)
40. Military doctrine of the Russian Federation. President of Russia, 2014. http://news.kremlin.ru/media/events/files/41d527556bec8deb3530.pdf
41. Tetrais B (2018) Russia's nuclear policy: worrying for the wrong reasons. Survival 60(2)
42. Chifu I, Nantoi O, Suchko O (2008) Societal Security in the trilateral Region Romania-Ukraine-Republic of Moldova/Securitate societală în regiunea trilateralei România-Ucraina-Republica Moldova, ediţie bilingvă. Editura Curtea Veche, Bucureşti. ISBN 978-973-1983-00-4
43. Chifu I (2023c) Rezilienţa în război şi caracteristicile intangibile. Cazul războiului de agresiune al Rusiei în Ucraina. Revista Infosfera. ISSN 2065-3395.
44. Chifu I (2022c) Securitatea, capacitatea geopolitică, rezilienţa societală şi politicile publice pentru a o pune în aplicare. Buletinul UNAp 11(3). https://revista.unap.ro/index.php/revista/article/view/1483
45. Chifu I (2018) Războiul hibrid şi rezilienţa societală. Planificarea apărării hibride, Revista Infosfera, februarie. ISSN 2065-3395
46. Chifu I (2021) Rezilienţa Strategică: de la stabilitate şi prevenţie la acţiune pro-activă şi adaptabilitate dinamică. Gândire Militară Românească
47. Chifu I (2022a) Lecţiile conflictului din Ucraina şi perspectiva rezilienţei societale. Revista Infosfera. ISSN 2065-3395
48. Chifu I (2022b) Rezilienţa contemporană: competitivitatea naţiunilor, capabilitatea statelor şi securitate societală. Revista Infosfera. ISSN 2065-3395.
49. Faulkner L, Brown K, Quinn T (2018) Analyzing community resilience as an emergent property of dynamic social-ecological systems. Ecol Soc 23(1). https://doi.org/10.5751/ES-09784-230124
50. Joint Research Centre (JRC) (2017) Building a scientific narrative towards a more resilient EU society. Part 1, A conceptual framework. Publications Office European Commission. https://data.europa.eu/doi/10.2760/635528
51. NATO(2016), Resilience, a core element of collective defense. https://www.nato.int/docu/Review/2016/Also-in-2016/nato-defence-cyber-resilience/EN/index.htm
52. Chifu I (2023b) Epistemological reassessment on the concept of resilience. Societal intangibles in the identity dilemma, Buletinul UNAp, nr. 4/2023. https://revista.unap.ro/index.php/revista/article/view/1790/1740
53. Chifu I, Simons G (2023) Rethinking warfare in the 21-st century. The influence and effects of the politics, information and communication mix. Cambridge University Press. ISBN:9781009355247
54. Chifu I (2024) Resilience, societal security and the intangibles of war. Ukraine's lessons learnt. Romanian Military Thinking, nr. 1/2024
55. Chifu I (2023a) Ameninţarea reală, a războiului de agresiune, de lungă durată,pe scară largă, în Europa. Adevărul. https://adevarul.ro/blogurile-adevarul/amenintarea-reala-a-razboiului-de-agresiune-de-2292769.html
56. NATO, *Secretary General* Annual Report 2021. https://www.nato.int/cps/en/natohq/opinions_193590.htm?selectedLocale=en

The Role of Strategic Communication in Relation to Security Culture and Societal Resilience. The Case of Ukraine

Adrian Lesenciuc

Abstract The triadic relation Strategic Communication—Security Culture—Societal Resilience is the optimal response to the conceptual blurring and to the fog of ambiguity that the hybrid confrontation entails. Although this triadic relation has been highlighted by numerous studies, most of them have usually highlighted only one of its component elements and have insisted on the role of institutional and instrumental preconfiguration. Strategic Communication (StratCom) is, however, a concept that addresses the cognitive level of the confrontation and assumes an organic or process approach and orchestral coordination (or directing) of information flows, irrespective of their nature. The case of Ukraine is the best example as regards the quasi-simultaneity of the institutional and organic implementation of the societal resilience, developed through the war. This article aims at highlighting the organicity of the aforementioned triadic design with an emphasis on strategic communication, in connection with the other layers of society's protection against hybrid threats, illustrated with the case of Ukrainian Strategic Communication and Societal Resilience.

1 Introduction

Association of such the concepts as Strategic Communication, Security Culture and Societal Resilience is not new, although all of them are relatively recent terms and cover significantly different areas in response to hybrid threats. Even if the terms were brought up together by different scholars, most often they emphasized only one concept, which coordinated the relation, rather than the relation itself in the dynamics of its functionality. For example, through the study *The Role of Strategic Communications in Enhancing Resilience and Stability in the EU's South-Eastern Neighborhood*, the researchers Katarina Kerysova, Michel Rademaker and Tim Sweijs highlighted the first concept, capable to: „(1) expose disinformation,

A. Lesenciuc (✉)
"Henri Coanda" Air Force Academy, Brasov, Romania

F.-C. Cirstoiu et al., *Medical Response Strategy in Case of Radiation Emergency Caused by the War in Ukraine*, NATO Science for Peace and Security Series B: Physics and Biophysics, https://doi.org/10.1007/978-94-024-2266-5_8

(2) improve media education and media environment, and (3) improve the EU's profile both inside the EU and in the EU's neighborhood" (Kertysova et al. [7], pp. 4–5). This concept was designed in the vicinity of the managerial tool proposed by the American military strategists, PMESII-PT, an acronym for the Political, Military, Economic, Social, Information, Infrastructure, Physical environment, and Time domains, which became indicators in the organization and adequate response to information in complex military exercises.

In the mentioned approach, the centrality of the StratCom concept also derives from the three-step design of the process aiming at of Societal Resilience, which has as its center of gravity the same term from the previously mentioned dynamic equation: I. Laying the Basis ((1) Develop a shared definition of StratCom, (2) Develop a StratCom mind-set, (3) Adopt a StratCom capability planning process, (4) Boost intelligence sharing), II. Enhancing Synergies (5) Create synergies, (6) Coordinate efforts more effectively, (7) Adopt a whole-of-government approach) and III. Mastering Strategic Communication (8) Strengthen the media environment, (9) Leverage new technologies, (10) Communicate a positive narrative, (11). Enhance resilience against unwanted interference (Kertysova et al. [7], pp. 16–18). If the three steps have been defined as (I) Strategic Communication, (II) Security Culture (and Education) and (III) Societal Resilience, which is consistent with our initial design, it would mean that the center of gravity is shifted from institutions to citizens/communities.

Moving the center of gravity from institutions to citizens/communities is, in fact, the stake of the debate starting from the concept of Societal Resilience, as the desired and final stage, designed, obviously, in relation to the ultimate objective of effectively combating hybrid threats. The literature proposes a series of works that partially change the perspective. Nikolov [16] emphasizes the resilience for effectively combating hybrid threats, starting from the institutional framework for defining societal resilience in accordance with NATO's strategic decisions (primarily during the Warsaw summit of July 8–9, 2016). Aatola and Juntunen [1] particularize the concept of societal resilience in accordance with a cultural frame of reference, the Nordic Model, analyzing "The 2017 Security Strategy for Science in Finland" and identifying the two major imprints on the Finnish security culture, the Nordic Model and Cold War and the need for an adequate response to the hybrid influence, with an emphasis on community flexibility in the response: "the flexibility in unpredictable situations and preparedness to respond to hybrid influence and cyber threats in their various forms and to enhance the needed capabilities".

Lesenciuc [13] focuses on hybrid threats and identifies the optimal response path through the mentioned triadic relationship, with an emphasis on the community component, Security Culture—Societal Resilience. The design of societal resilience in the case of the aforementioned works does not differ substantially and involves the active engagement of individual and institutional actors, state and private, in a cohesive and coherent framework:

Resilience is an essential basis for credible deterrence and effective fulfillment of the Alliance's core tasks. Under the North Atlantic Treaty, all Allies are committed to building resilience. In today's security environment, resilience more than ever requires a full range of capabilities, military and civilian, and active cooperation across governments and with the private sector. It also requires engagement with partner countries and other international bodies, and continuously updated situational awareness (Nikolov [16], pp. 91–92).

One of the important works, which brings into the debate, starting from the mentioned ternary structure, yet without mentioning the causal relationships between the terms, the current context of applicability, the war in Ukraine, is that of Olga Reznikova, who builds the concept of societal resilience in accordance with an explicit theoretical framework. The work starts from systems theory perspective, continuing with the intermediate level of the interdisciplinary resilience concept, able to design the analytical framework at the level of security studies. Through the comparative analysis of the conceptual approaches to building societal resilience in NATO: "Adapt to uncertainty; timely identification and elimination of vulnerabilities; development of respective capabilities and interaction", EU: "Providing:—ability of Member States and the whole Union to confront the full spectrum of threats without prejudice to common democratic values;—trust to institutions;—sustainable development;—good governance;—ability to reform", UN: "Enhance states' crisis management, develop risk management systems, enhance civil protection and interaction between the government and population, enhance resilience of energy, water, food supply, transport systems, etc." and OECD: "Enhance risk assessment and organize the respective efforts on the basis of wide cooperation; identify vulnerabilities of states and their societies; support in development of national strategic and program documents with respect to the national resilience enhancement, etc" (Reznikova [17], pp. 222–223) and through the relational analysis of structures holding the role of ensuring societal resilience in Ukraine and inter-agency cooperation, the paper highlights the need for systemic approach and measures to ensure functionality in particular areas of application (Reznikova [17], p. 348), which, regardless of the involvement of the institutions and the coherence of the measures, depends on the level of the nation's security culture, implicitly on that of each citizen: "Important for ensuring national resilience in Ukraine is the formation of a culture of security and resilience in the state and in society" (Reznikova [17], p. 422).

In the light of these results, through this article we aim to highlight the role of Strategic Communication as a dynamic process that shapes the functional relations with the other elements of the structure, Security Culture and Societal Resilience, shifting the focus from the explicit institutional design to the natural exercise of resilience from the citizen's perspective.

2 Strategic Communication as a Holistic Approach on a Cognitive Level

The analysis of the Strategic Communication—Security Culture—Societal Resilience triad starts from the need to design the appropriate response to hybrid threats. This response is consistent with the domain in which hostilities are conducted and in which the effects are intended, more precisely the cognitive domain. This connection with an area—the cognitive one—not yet included in the warfighting domains neither in the American nor in the NATO doctrines, raises a series of debates and requires precise conceptual clarifications. Starting from the definition of Strategic Communication proposed by the NATO Strategic Communication Centre of Excellence in Riga, "a holistic approach to communication based on values and interests that encompasses everything an actor does to achieve objectives in a contested environment" (Bolt [3], p. 31), Elīna Lange-Ionatamišvili, senior StratCom expert of COE in Riga, highlights this aspect:

This definition implies that the cognitive domain is constantly being contested by a variety of actors. It also conveys that strategic communications sits at the core of policy-making and implementation. It is a mindset that allows to conduct statecraft through the projection of values-based narratives and implementing policies through aligned words and actions to ensure the consolidation of democracy, appropriate levels of national security, and cohesion and welfare of the whole of society. The problem of hybrid attacks lies within their ambiguity and multi-faceted character, complicating attribution and response. Often it is not possible to determine the origin of a hybrid attack or to instantly grasp the scale of the attack across different domains. A strong strategic communications approach means decreased vulnerabilities of the country as a whole, and can be seen as a pre-emptive action to hybrid attacks (Lange-Ionatamišvili [12], pp. 83–84).

Probably the most important response attempt to the hybrid threats on the informational level consists of understanding the cognitive domain as a warfighting domain, starting, in the simplest way of analyzing, from the (doctrinal and theoretical) definition of the informational environment. The warfighting domains in the Alliance's design are: air, land, sea, space, and cyber—even if, for operational reasons, NATO proposed the Joint Air and Space Power concept, developed by Joint Air Power Competence Center (JAPCC) in Kalkar, Germany –, but precisely within the definition proposed by the Allied Command Transformation with which JAPCC operates, the objective of the multi-domain operations aims at influencing the attitudes and behaviors of all adversaries and its allies, i.e. targeting within the (yet) undefined cognitive domain:

NATO's Military Instrument of Power enables, prepares, plans, orchestrates, and executes synchronized activities, across all domains and environments, at scale and the speed of relevance, in collaboration with other instruments of power, partners and stakeholders. This delivers tailored options at a time and place of our choosing that build advantage, across the shaping, contesting and fighting contexts, in order to decisively influence the attitudes and behaviors of adversaries and connected audiences.

Multi-Domain Operations aim to orchestrate and synchronize military activities with non-military activities, across all domains and environments, to enable commanders to deliver converging effects. (Allied Command Transformation, *Multi-Domain Operations Concept*, Dec. 2021, *apud* JAPCC [5]).

It is probably the problem of ambiguity that hybrid threats assume and exploit that causes circumspection in the definition of a concept, Strategic Communication, in direct correlation with the cognitive domain, but defined outside of it, as long as the domain itself is neither even defined nor unambiguous in use. Not defining the cognitive domain, however, does not eliminate the problem of ambiguity, especially since: (1) the ultimate goal of any war is the defeat of the opponent's will to fight; (2) the field in which this goal is achieved is the cognitive field and (3) in defining the informational environment—on the basis of which the cyber field was established as an independent field—the cognitive dimension cannot be ignored. The information environment (IE) is defined and understood within NATO as "the cognitive, virtual and physical space" (MC 422/6, NATO Military Policy on Information Operations, Annex B), in which information is received, processed and transmitted. This definition was taken from the previous editions of this document and did not involve any stage where the cognitive dimension of the information environment was not taken into account. This design is based on the American doctrine concept of information environment, which involves three distinct but interrelated dimensions, and which has been defined identically in the various editions of the American doctrine of information operations and in its related documents. Different perspectives on the information environment and its dimensions do not alter the understanding of the concept, but require a more appropriate approach depending on the angle from which the process is viewed:

The transmission-centric model deconstructs the IE into three separate dimensions through which data flows. The description of the informational dimension is also part of the description of the other two dimensions so it becomes difficult to distinguish. The informational dimension is where we collect, process, store, disseminate, and protect information. However, humans and computers perform these five functions in both the physical and cognitive dimensions. This construct works well in analyzing how data flows through information systems and networks to reach a receiver, but becomes problematic when trying to understand the meaning activities communicate in a pervasive and dynamic IE. Because this model focuses on the flow of information, it separates the mind (cognitive) from the body (physical), and the thoughts (informational) from the mind (cognitive). (*JCOIE* [6], p. 3).

However we consider the information environment, we cannot ignore the cognitive dimension of the human-centric approach, because, ultimately, this is the area where the decisive battle of any war is fought. That is why, returning to the definition of StratCom, it is important that the concept be applied on all levels, from the strategic one—in terms of B.H. Liddell Hart, of the Grand Strategy—up to the tactical one, assuming a "whole-of-government approach" and a national coordinated effort "to develop and strengthen itself politically and militarily", "designed to target different domains simultaneously" (Lange-Ionatamišvili [12], pp. 83–84).

3 Strategic Communication as an Umbrella-Concept for Military Communication/Information Flows

The coherence of strategic communication must be in use in parallel, in the political and social domains, but also in the military, which means coherence from the executive level to the level of small communities, respectively from the military strategic level (Small Strategy) to the tactical level. The focus is on the rigorous organization at the political and social level, and from this explicit requirement also follows the need for citizens to be aware of hybrid threats (security culture) and to respond appropriately to them (societal resilience), as also results from The Hague Centre for Strategic Studies guide regarding the optimal response to this type of threat:

Strategic communication must be synchronized across all branches of government, rather than inconsistently depending on the military, and control over the narrative and associated messaging '[...] must be factored in from the start' and understood as an 'organizational responsibility'. StratCom must not be limited to press offices or the publication of official statements, doctrines, and strategies: governments should invest in disseminating strategic messages to boost a coherent and credible posture in line with the guiding principles and values pursued. As an overarching tool that informs the overall strategy, effective StratCom must be targeted towards one's population and key stakeholders (such as companies, organizations, etc.) to raise societal awareness and enhance resilience. It must target like-minded partners through diplomatic, legal, military, and economic cooperation to gain strategic support and boost unity of intent at the international level. Finally, it is crucial that the message effectively targets the opponent engaging in hybrid operations. (Bertolini et al., [2], p. 32).

Coherent communication in the conditions of hybrid threats, coordinated through StratCom, requires taking into account the information environment defined within the military doctrines of information operations. This design induces two possible paths to misinterpretation: (1) understanding StratCom as an eminently military function and (2) considering the military information infrastructure as unitary. If in the former case the specialized literature is explicit, in the latter, things are more complex and based on the two fundamental purposes of communication as a whole: to inform and to maintain open communication channels, respectively to influence.

Information communication is carried out with the aim of obtaining and maintaining the necessary level of confidence of the internal and external public of the military organization in it and is carried out through public relations (PR)—public affairs (PA) in the American doctrine or media Operations (Media Ops) in the British one—respectively through civil military (co)operations (CIMIC), the one for keeping communication channels open through public diplomacy and military diplomacy, while influencing communication is carried out through information operations (Info Ops), mainly psychological operations (Psyops), cyber network operations (CNO) and military deception (Mildec). StratCom coordinates all these communication flows, which are distinct and non-interfering. The most important flows are those built on the relationship of trust—the only effective social link—between institutions and citizens, achieved through information communication, mainly through public relations/ public affairs/ media operations, based on 'Tell the Truth!' principle, respectively on undermining the confidence, will, understanding and action capacity of the adversaries, through information operations (which include psychological operations). Basically, both flows operate on a cognitive level, which thus could be extended from the meaning of a simple component of the information environment to the warfighting domain. Summarizing the two designs, joined by with definitions from the old doctrines, the allied and the Romanian ones, on information operations, can be illustrated as not only different, but also antagonistic in terms of implementation.

The two communication flows should not come into direct contact, because influence communication can corrupt information and, implicitly, can lead to a dramatic decrease in the level of external public trust in the military organization. The two flows must be placed under the umbrella of the integrative concept, StratCom, which includes Public Diplomacy, Public Affairs (PA), Military Public Affairs, Information Operations (Info OPs) and Psychological Operations (PSYOPS)—starting with the document PO (2009)0141, NATO Strategic Communication Policy of 2009 (*apud* NATO Strategic Communications Handbook, 2015: 5)—which aims to coordinate and synchronize all communication activities, as summarized in some synthesis works: "Strategic Communication (StratCom) is recognized by NATO as a concept that brings to the same denominator all the civilian and military activities carried out to significantly impact on the information environment" (Vasile [18], p. 47). Strategic communication serves, therefore, "to make order in the inflation of concepts, grafted on the communication processes with different audiences" (Vasile [18], p. 56) and to bring the communication flows to the same common denominator as an integrated approach, based on values, principles and interests. From this perspective, StratCom can be viewed as a process rather than as a function or a distinct capability (although most often it is seen as a function and possibly as a capability associated with the cognitive level effects), which contributes to a better understanding of the ternary relationship Strategic Communication—Security Culture—Societal Resilience from a dynamic perspective, and in which the focus is not on communication itself (on its semantic and syntactic dimensions), but on maximizing the effects of communication, i.e. on the pragmatic dimension, expressed by the metaphor of the orchestra:

The character of each functional area is significantly different; they all operate in specific surroundings, bringing different sensitivities in the communication work. The idea of StratCom being a process is admitting that orchestra conductor doesn't have to know how to play on each and single instrument (Łyko [15], p. 100).

The orchestral communication in its strategic meaning coordinates the harmonics of each communication flow, and as regards the coordination of the two military flows already mentioned, information and influence, StratCom coordinates the process of integrating functions and capabilities with other military activities with the aim of streamlining actions on a cognitive level.

Regardless of the communication actions, StratCom ensures cooperation and coordination of flows and provides the strategic perspective on the optimal use of communication architecture, both in peacetime and in wartime. Moreover, StratCom should be a source of clarification needed for effective communication under the conditions of the blurring of meanings associated with the terms peace and war, then with the ambiguity of the concept of truth, i.e. with the imposition of a form of hybridization of life, including the post-truth perspective, which coagulates in the hybrid expression of different types of threats, up to hybrid conflictuality.

4 Conclusions

The conflict in Ukraine did not surprise the host state unprepared from Strategic Communication perspective. If in the year of 2024 there is a practical guide to the implementation of StratCom concept in Ukraine under the conditions of war [9], this was possible due to the fact that there existed a prior preparation and adaptation of the functional concepts of NATO in the current practice of military operations, both kinetic and non-kinetic, in which Ukraine was involved. Not by chance, in the case of this guide the most consistent chapter (from a qualitative point of view) is the one regarding the StratCom approach as a proactive response to Russian hybrid threats, and the most consistent one from a quantitative perspective is that regarding the good practices that understand "StratCom as a system" (chapter subtitled "A unique experience of the institutions of the security and defense sector in Ukraine in the first Year of the War"). Prior to the outbreak of the war, there were, regarding the dialogue between NATO and Ukraine, a series of Joint Working Groups, one of which leading to the emergence of a Partnership Program on Strategic Communication (in 2015), and starting from these training frameworks, the National Academy of the Security Service of Ukraine developed an interdepartmental platform entitled "Strategic Communications in the Context of Hybrid War", designing and proposing for implementation the principle "single voice", then the principle "Tell the Truth!"—implicitly the necessary trust, the associated truth and the adjacent national values; in a previous article, the same researcher, L. Kompantseva, proposed the "formation of a scientific and practical discourse of StratCom based on the concepts "national values," "truth," "trust" (Kompantseva [8], p. 74). Kostyrev [11] considers that the implementation of the StratCom comcept in Ukraine involved in two stages: a first stage, from 1991 to 2013, of the implementation of the Master Plan regarding NATO-Ukraine Strategic Communication, coinciding with a gradual, but slow, implementation of the *forms*—Kotsyrev refers explicitly to the reform of institution and enforcement instruments—followed by a second stage, from 2014 to the present, which involved an adaptation of the *substance,* as long as the most important issue in enforcement is the very understanding of the purpose of communication (including strategic) and the related principles [11]. The second stage basically has two phases: 2014–2022 (February), and from the beginning of the Russian invasion to the present.

From the organic perspective on StratCom concept, and on the triadic structure of Strategic Communication, Security Culture and Societal Resilience, the second phase of the second stage of implementation involves acceleration and adaptation of NATO conceptual and doctrinal forms to the Ukrainian substance, due to the societal resilience acquired through the exercise of the conflict that broke out in Eastern Ukraine in 2014. Societal resilience in Ukraine involved strengthening antibodies during illness. Two years ago, we identified, through the study of Boulègue et al. [4] that the *Revolution of Dignity* started in 2014 and this is the starting point in considering the Ukrainian societal resilience:

Ukraine can be viewed as a political 'laboratory' in which Russia has tested a variety of measures to exert influence, and at the same time as an example of resilience. Since 2014, many of the levers of Russian influence have weakened as a consequence of civil society mobilization associated with Ukraine's 'Revolution of Dignity' and subsequent reforms. (Boulègue et al. [4], p. 2).

Through the voice of civil society organizations, then through the awareness of the need for an adequate response to the Russian hybrid challenges, Ukraine began to develop a series of policies in order "to adapt to the uncertain security environment and to recover from the violent military actions in the east of the country" (Lesenciuc et al. [14], p. 29). We found in article that societal resilience can be configured in two formative contexts: through education and security culture or through war. Arguing for the first model, we also note that the second way is applicable to Ukraine and Israel and that, in the case of Ukraine, the Korostelina model [10, p. 20] summarizes—based on previous analytical frameworks considered by the researcher—the forms of manifestation of soft power that led to the formation of a strong civic voice, expressed in parallel with strong government support, consisting of the promotion of national identity and through other measures, including a strong anti-corruption campaign.

Perhaps this societal resilience is the element that Putin did not take into account when he ordered the invasion, hoping that he would solve the problem of his own policy imposition in a few days. The Ukrainian response was based on the two previously mentioned elements, a set of forms—StratCom—implemented over time, which made it possible to control without major syncopations of strategic communication and the implementation of related principles, equally proposed by scientific researchers and practitioners, and a substance—societal resilience—already adapted to the new forms, trained during the 8 years of war: "Societal resilience in Ukraine is the result of the common learning of the Ukrainian community to respond to a threat that has not ceased to be evident for eight years, until the invasion. Under these conditions, Ukraine's resilience is driven by the crisis and conflict in the post-Maidan period" (Lesenciuc et al., [14], p. 30), which allowed information flows in war conditions not to be altered by the ambiguity of the Russian hybrid threats.

Only the dynamic understanding of StratCom in the orchestral or process perspective of communication allows the transformation of societal resilience into the most formidable defense. The real lesson of Ukraine is not that of a model of societal resilience, developed by (through) war, but that societal resilience based on already implemented institutions and tools of Strategic Communications, that truly turned into the power capable of stopping the aging war machine of the Russian Federation. Therefore, due to the substance (background) adapted (even due to the war started in 2014) in parallel with the adopted forms, Ukraine produces an important lesson through the triad of Strategic Communication, Security Culture and Societal Resilience in what the appropriate response to hybrid threats entails.

References

1. Aaltola M, Juntunen T (2018) Nordic model meets resilience—Finnish strategy for societal security. In: Aaltola M, Kuznetsov B, Sprūnds A, Vizgunova E (eds) Societal security in the Baltic sea region. Expertise mapping and raising policy relevance, Riga, Latvian Institute of International Affairs, pp 26–42
2. Bertolini M, Minicozzi R, Sweijs T (2023) Ten guidelines for dealing with hybrid threats. A policy response framework. The Hague Centre for Strategic Studies, Hague
3. Bolt N, Haiden L (2019) Improving NATO strategic communications terminology. NATO Strategic Communication Centre of Excellence, Riga
4. Boulègue M, Lutsevych O, Marin A (2018) Civil society under Russia's threat: building resilience in Ukraine, Belarus and Moldova. Chatham House. The Royal Institute of International Affairs, London
5. Joint Air Power Competence Center (JAPCC) (2021) Multi-domain operations. JAPCC. https://www.japcc.org/warfare-areas/mdo/. Accessed 11 Mar 2024
6. Joint Chiefs of Staff (2018) Joint concept for operating in the information environment (JCOIE). US Department of Defense, Washington, DC
7. Kertysova K, Rademaker M, Sweijs T (2022) The role of strategic communications in enhancing resilience and stability in the EU's South-Eastern Neighborhood. The Hague Centre for Strategic Studies (HCSS), Hague
8. Kompantseva L (2021) Research, education, and practice of StratCom in the security Service of Ukraine in interagency settings. Inform Secur 48(1):68–93
9. Kompantseva L (ed) (2024) Implementation of strategic communication under the war conditions, Recommendations and best cases. Practical guide. National Academy of Security Service of Ukraine, Kiev
10. Korostelina KV (2020) National resilience to protracted violence in Ukraine. Peace Conflict Stud J 27(2). https://doi.org/10.46743/1082-7307/2020.1689
11. Kostyrev AG (2023) NATO-Ukraine strategic communications: theory and practice. Гілея: науковий вісник. Київ: «Видавництво «Гілея» 1–2(180–181):82–95
12. Lange-Ionatamišvili E (2022) The role of strategic communications in preventing hybrid threats in the Baltic States. In: Sprūds A, Bērziņa-Čerenkova UA, Broka S (eds) Hybrid threats in Baltics and Taiwan: commonalities, risks and lessons for small democracies. Latvian Institute of International Affairs, Riga, pp 80–95
13. Lesenciuc A (2023) Războiul hibrid sau Întoarcerea prin disimulare doctrinară la războiul absolut. Foreword by Antonia Colibășanu. CTEA, Bucharest
14. Lesenciuc A, Nagy D, Lesenciuc S (2022) Societal resilience. Between resilience through education and resilience through war. Redefining Community in Intercultural Context 10(1): 25–31
15. Łyko G (2017) Strategic communication in a changing security environment. Zagadnienia Społeczne 1(7):95–110
16. Nikolov O (2018) Building societal resielience against hybrid threats. Inform Secur Int J 39(1): 91–109
17. Reznikova O (2022) National resilience in a changing security environment. Monograph. National Institute for Strategic Studies, Kiev
18. Vasile V (2017) Comunicarea strategică. Repere conceptuale și evolutive (I). Gândirea militară românească 2:47–60

Risks of Nuclear Threats in Modern War

Andriy Stavytskyy

1 Introduction

The war in Ukraine has become one of the largest military conflicts since World War II. It has heightened concerns about the potential for either controlled or uncontrolled radiation contamination over a wide area. Currently, it can be stated that the war is fought primarily with the use of conventional weapons. Despite the calls of certain journalists and semi-officials in the Russian Federation to launch nuclear strikes on Ukraine or NATO countries, no movements towards such actions have been observed so far. Furthermore, it is important to state that the use of nuclear weapons in general will not solve the underlying causes of the conflict, therefore, the probability of their use appears to be quite low in this particular war.

However, this does not mean that risks of both spontaneous use of nuclear weapons and damage to facilities containing nuclear fuel cannot arise. Moreover, the geopolitical situation in the world does not exclude new conflicts at all. That means the possibility of both the use of nuclear weapons and radioactive contamination due to damage to certain critical infrastructure. Therefore, it's important to investigate these risks and their implications in greater detail about the war in Ukraine. However, similar considerations can be made regarding future conflicts.

The possibility of using nuclear weapons has been debated since the start of a full-scale invasion in 2022. However, in March 2024, it intensified. Due to the largest monthly losses since the beginning of the war, the officials of the aggressor country began to openly indicate the possibility of a nuclear strike without hesitation. At the same time, Western countries began to recognize the presence of their soldiers on the territory of Ukraine. Of course, under such conditions, the possibility of a direct clash between the Russian and NATO troops increases, which can stimulate the

A. Stavytskyy (✉)
Taras Shevchenko National University of Kyiv, Kyiv, Ukraine

F.-C. Cirstoiu et al., *Medical Response Strategy in Case of Radiation Emergency Caused by the War in Ukraine*, NATO Science for Peace and Security Series B: Physics and Biophysics, https://doi.org/10.1007/978-94-024-2266-5_9

escalation towards a nuclear disaster. In any case, the aggressor country is considering the possibility of using at least tactical nuclear weapons, which means that it is necessary to clearly understand the devastating consequences of such actions for Ukraine and the world. The purpose of this paper is to analyse potential options for deploying nuclear weapons during the Russian-Ukrainian war and the possible consequences of such actions.

2 Literature Review

For a long time, waging a full-scale war between the main countries of the world was considered unrealistic precisely because of the possibility of using nuclear weapons. However, after the start of a full-scale Russian-Ukrainian war in 2022, the rhetoric of both officials and experts largely changed. In particular, Russia began to consider the possibility of a tactical nuclear attack on Ukrainian troops [1]. It is obvious that such an attack carries not only war risks but also threatens nuclear stability throughout the planet. After a nuclear weapon is asserted against a country that does not have such weapons, the interest of almost all countries in obtaining nuclear weapons will increase significantly, which will sooner or later lead to a nuclear conflict.

Moreover, it can begin both with a direct attack on another state and with a mistaken reaction to the actions of another nation [2], which is connected with the growing tension between different blocs in the world. Although currently, the probability of starting a nuclear conflict is rather small [3], the situation can change at almost any moment. For example, parts of missiles and UAVs launched by Russia have repeatedly fallen on the territory of Romania and Poland. If there are casualties as a result of such hits, it will be very difficult for NATO countries not to intervene in the war. There have even been simulations of such a war quickly escalating into an exchange of nuclear strikes [4].

At the same time, the start of a nuclear war without the participation of NATO countries in the conflict should be recognized as a very unlikely, but not impossible, event ([5]: 342). It is obvious that even a regional nuclear war will have numerous consequences, in addition to the direct death of military personnel and civilians in the city of impact. Among the main risks, the possibility of famine, the impossibility of globally providing water and food supplies, the pollution of the territory and the increase of population and animal diseases, climate change and many other aspects are often highlighted.

For example, M. Bivens [11] estimated how catastrophic the scenario of massive use of nuclear weapons would be due to the direct death of people and the onset of famine in 2 years. According to his estimates, about 260 million people will die if 100 warheads of 15 kt are used, and about 2.5 billion people will die if 500 warheads of 100 kt are used. Unfortunately, such a scenario proves only one thing: there is no way to restore the planet after a nuclear war, so the only way to make it impossible is to completely ban such weapons. However, practice shows that any prohibitions are circumvented over time, and therefore the risk of nuclear escalation in the near future cannot be ruled out. This thesis is confirmed by the speeches of the President of the Russian Federation, who announced the complete renewal of nuclear weapons and the lack of hesitation to use them.

According to R. Savolainen [6], the risk of war can be reduced by increasing the exchange of information at all levels. Therefore, the purpose of this work is to analyse the possible consequences of a nuclear attack on the territory of Ukraine.

3 Results

The Russian Federation is constantly engaged in the modernization of its nuclear forces. At the same time, the number of deployed strategic warheads is decreasing, but the total number of nuclear missiles is quite significant. In particular, as of the end of 2021, it had 6257 missiles, of which 1760 were prepared for decommissioning. At the same time, there were at least 1444 strategic warheads deployed [7]. The strategic missile forces of Russia are currently deploying several variants of mine and mobile missiles. Mine-based missiles include SS-18, SS-19, SS-27 Mod 1, SS-27 Mod 2, and mobile ICBMs include SS-25, SS-27 Mod 1 and SS-27 Mod 2. According to official information, these missiles are 95 per cent of Russia's strategic missile forces.

The Use of Strategic Nuclear Weapons One of the most significant risks in modern warfare is the use of strategic nuclear weapons. These weapons are designed to cause mass destruction and would lead to the loss of hundreds of thousands of lives and extensive radiation contamination over a large area. This risk is particularly alarming given the advanced capabilities of modern nuclear arsenals. At the same time, this weapon is aimed primarily at the destruction of entire cities and has significant pollution. From the point of view of strategy, it is not expedient to strike a similar attack on one city, to solve the problem of destroying the country, it is much more effective to destroy the capital itself.

However, it should be clearly understood that unlike in the middle of the last century, modern military command posts are sufficiently decentralized, and therefore even the destruction of the capital or another large city is not a guarantee of the collapse of the army in defence. However, the consequences will be extremely severe in terms of the number of civilian casualties, contamination of the territory, and destruction of critical infrastructure facilities.

When deciding to launch a nuclear attack, the most likely missiles will be Iskander-M or Kinjal, the warhead is estimated to have a yield of 5–50 kilotons of TNT. Of course, the Russian Federation can decide to strike Ukraine with an intercontinental ballistic missile, such as Topol, Yars, R-36 M2 Voevoda, or from underwater missile carriers. But such a launch of even one ballistic missile will be immediately detected by the USA and, with a fairly high degree of probability, may already lead to a mass launch of American missiles.

A nuclear strike would produce a blinding flash of light and a powerful heat wave that would spread instantly away from the epicentre. The diameter of these zones can reach more than 10 km. If you look at the map of Kyiv, which has a length of about 40 km, then such an explosion will not lead to complete destruction.

The maximum size of the nuclear fireball could be expected in a radius of 230 m., a radiation radius of 1200 m, in which individuals will receive a lethal dose of radiation, or 15% of survivors will eventually die of cancer as a result of exposure. Within a radius of 2100 m most residential buildings collapse, injuries are universal, fatalities are widespread. The chances of a fire starting in commercial and residential damage are high, and buildings so damaged are at high risk of spreading fire. Within a radius of 2400 m third-degree burns extend through the layers of skin and are often painless because they destroy the pain nerves. The total damage radius will be about 6 km, in which there will be destruction of infrastructure and injuries, primarily burns. It is expected that 50,000–60,000 people will die instantly, and the number of wounded will reach 250,000. However, after the strike, the risks of new losses among the population will be even higher.

As the beginning of the war in Ukraine and the attacks on critical infrastructure showed, in the event of panic, power outages, and destruction of roads and bridges, the opportunity for adequate assistance to the population and provision of medicine, water, and food is actually lost. In such conditions, a complete paralysis of supply to the relevant city and significant problems with evacuation are expected. This means that most of the wounded will not receive help and will be doomed to die. Thus, under such a blow, the immediate or delayed death of up to 10% of the capital's population is expected. When simulating a similar attack on other large cities of Ukraine, the situation is even a little bit worse: 30% of the population of Lviv, 20% of Dnipro, 15% of Kharkiv, and 10% Odesa will be destroyed.

It should also be noted that there is a non-zero probability of interception of nuclear missiles by air defence systems. The destruction of a missile with a nuclear warhead in the air should not lead to a nuclear explosion, but fragments of the charge itself will be scattered over a certain area, which will require measures to find and deactivate them.

Tactical Nuclear Weapons In contrast, tactical nuclear weapons have a more limited radius of effect. This means that while their use would not result in the same level of massive casualties as strategic weapons, they still pose a grave threat. The reduced density of troops in modern warfare might mitigate casualties that some extent, but the localized radiation contamination remains a significant concern. However, it should be understood that tactical nuclear weapons were developed after the Second World War precisely to create conditions for a breakthrough in the enemy's saturated defences when several dozen divisions are opposing each other in certain areas of the front. In the current war, such a density of troops is unrealistic. The use of Iskander missiles of lower power, for example, 10 kilotons in rural areas of the Zaporizhia region, as a tactical nuclear weapon, will mean the guaranteed death of a few thousand people, which obviously will not fundamentally change the situation at the front. Thus, the use of tactical nuclear weapons in a non-massive version will not lead to significant changes and is therefore impractical. The use of strategic nuclear weapons also seems unlikely because of the risk of nuclear war with the United States.

On the other hand, the use of tactical nuclear weapons can cause immense damage to the Ukrainian economy and global food security. According to data for 2023 [8], 45.2% of Ukrainian exports are products of the agro-industrial complex, which after the hypothetical use of nuclear weapons will not be in demand in the world due to radiation protection. This means a devastating blow to the entire Ukrainian economy, mass unemployment among farmers, and the absence of the possibility of restoring their business even in the medium-term perspective. The pollution of resources and territory, although it will not have catastrophic physical consequences, will have devastating psychological consequences for other countries that will impose a ban on the supply of goods from Ukraine. An obvious ban on transporting goods through European territory will be an equally painful blow, which will again be caused by protection against the spread of radiation. Thus, although a tactical nuclear strike cannot significantly change the distribution of military forces, it can undermine the economic and psychological foundations of the Ukrainian resistance.

Accidental or Non-accidental Hits on Nuclear Facilities A third risk involves the accidental or deliberate targeting of nuclear power plants. While the probability of a direct hit on a nuclear facility may be relatively low, the consequences of even a partial destruction of such a facility could be catastrophic, similar to the Chornobyl accident in 1986. In contrast, nuclear power plants in a state of cold shutdown are less vulnerable to such attacks. At the moment, there are several nuclear power plants operating in Ukraine: Rivne, Khmelnytska, and South-Ukrainian. The fourth station (Zaporizhia) is occupied and taken out of actual operation, being in a state of cold shutdown. The Russian Federation regularly bombards Khmelnytskyi and Rivne, but there has been no incident with nuclear plants.

On September 19, 2022, the Russian army struck the South Ukrainian NPP with an Iskander missile. A powerful explosion occurred just 300 meters from the. The NPP building was damaged by the shock wave, and more than 100 windows were broken. One of the hydraulic units shut down. Three high-voltage power lines were also shut down.

Although the Russian Federation disabled the transformer stations leading from the NPP, there was no immediate threat of a direct hit yet. However, this does not rule out that the consequences of air defence operations may lead to debris falling onto the territory of nuclear plants. At the same time, the strength of reactor structures must withstand even a direct hit by missiles.

Attacks of Scientific Laboratories Another risk factor is the potential targeting of scientific laboratories where experiments involving uranium or other radioactive materials are conducted. Fortunately, Ukrainian laboratories have not held significant volumes of nuclear fuel, so the damage would probably be localized. Similar risks were observed at the Kharkiv universities, in which scientific research was actively conducted. For example, since the beginning of the war, Russian troops have repeatedly shelled the Kharkiv Physical and Technical Institute. The experimental nuclear facility "Neutron Source" is located there. On March 17, the nuclear research facility in Kharkiv was cut off due to shelling by the Russian Federation.

However, even with a direct hit and starting a chain reaction, the power of the explosion should not exceed several kg in TNT equivalent, which will only lead to local destruction. At the same time, the level of pollution can be relatively easily localized, and the existing capacities of rescue services will be able to carry out a timely evacuation. Moreover, it should be noted that the absolute majority of such laboratories are taken from the frontline areas, and therefore the probability of such a scenario is also quite low.

It is important to note that, at present, war activities primarily involve conventional means. The probability of a transition to a full-scale nuclear scenario is relatively low. On the other hand, the analysis of historical events shows that the current war will definitely not be the last on the European continent, if such an entity as Russia, which constantly starts or gets involved in new wars, remains.

We should not forget other problems that are currently observed in Ukraine. In particular, we are talking about obvious problems with the supply of weapons and ammunition from the allies (for example, only a little more than a third of the million shells promised for 2023 were delivered), and delays in funding from both the EU and the USA. As a result, the counteroffensive of Ukrainian troops planned for 2023 failed. For example, due to delays in the delivery of military equipment, 4 brigades were completely unprepared for a counteroffensive [9].

All these actions led to Ukraine's loss of important towns of Maryanka, Avdiyivka, Bakhmut and others. Although this is not yet a strategic or operational failure, it significantly lowers the morale of both the army and the population. This leads to the need for a significant rotation in the Ukrainian army, the next wave of mobilization, which in 2024 should include almost 500,000 men. Of course, such decisions are difficult and unpopular, undermine the Ukrainian economy and reduce opportunities for defence. If the situation worsens, the capitulation of Ukraine will lead not only to the capture of its entire territory but also to Moldova, in which the self-proclaimed Transnistrian Republic is acting. Given the exhaustion of military resources in Europe, we should expect an increase in Russia's expansionist desires, which will make a direct conflict between the Russian Federation and NATO inevitable. Given the incomparability of military potentials, the exchange of nuclear strikes in this particular conflict is more likely. This is because the strikes will be carried out at a greater distance from the Russian Federation, the readiness to respond with a nuclear strike is doubtful, and the fact of the strike may demoralize the population of Europe.

Considering these threats, Europe must consider a strategic response to radiation threats. One approach could be to advocate for the division of Russia into multiple states (appr. 7–12 states), with at least half of them adhering to democratic values and being disarmed of nuclear weapons. This approach aims to reduce the potential for nuclear conflict in the region.

Should efforts to divide Russia and promote disarmament fail, European countries must prepare for the possibility of a nuclear conflict within the next 10–15 years. This preparation should include a comprehensive review of the medical response and mass emergency resolution system, which demonstrated its inadequacy during the 2020 pandemic. Among the reforms required, one crucial factor is the autonomy of all medical institutions. Additionally, there should be a systematic distribution of protection equipment and medicines that ensure the readiness of healthcare facilities in the event of nuclear emergencies. The next possibility is the wide implementation of autonomous digital diagnostic systems based on artificial intelligence, which should help not only in real-time with the treatment of people but also become the basis of treatment in the event of a nuclear explosion.

4 Conclusions

Russia's invasion of Ukraine raised the risk of a nuclear conflict on the planet to the highest level since the Caribbean crisis [10]. Although the war is currently being waged primarily with conventional weapons, there is a constant risk of escalation leading to at least a tactical nuclear strike. However, even an attempt to make such a strike could lead to the expected response from the Allies and the start of a world nuclear war. Currently, concerns exist that the Russian Federation might be the only country willing to use nuclear weapons first. That is why the world community should increase its political and diplomatic influence on this country to prevent the use of nuclear weapons under any circumstances.

At the same time, it should be understood that Putin's regime has already committed a huge number of war crimes, completely usurped power in the country, and therefore has no opportunity to make concessions because this could lead to the fall of power in the country. A certain attempted coup by Eu. Prigozhin in 2023 showed that, apart from militaristic factors, there are no other factors to preserve the regime. For this reason, the existing powers will try to maintain power at any cost, including using nuclear leverage. The only more or less real way out of this situation can be a significant military defeat of Russia on the battlefield, which will significantly reduce support for the regime and increase the likelihood of internal coups. Any further weakness of the new regime should be used to force the country to abandon its nuclear weapons and to divide the country to ensure long-term European security.

The risks associated with nuclear threats in modern warfare are complex and multifaceted. The lessons learned from the conflict in Ukraine should serve as a stark reminder of the potential consequences. It is our responsibility to advocate for peace, promote disarmament, and be prepared for any eventuality that safeguards the lives and well-being of our nations.

References

1. Ramzy N (2022) The existential threat of nuclear war: a psychoanalytic comment. Int J Appl Psychoanal Stud 19(1):3–15
2. Baum S (2022) How to evaluate the risk of nuclear war. BBC Future. 11 March 2022. Available at https://www.bbc.com/future/article/20220309-how-to-evaluate-the-risk-of-nuclear-war
3. Horowitz L, Wachs L (2022) Russia's nuclear threats in the war against Ukraine. SWP comment 2022/C 29, 20.4.2022. Available at https://www.swp-berlin.org/en/publication/russias-nuclear-threats-in-the-war-against-ukraine
4. Tiron R (2022) Preventing nuclear 'Armageddon' hinges on US policy of ambiguity. Bloomberg Government, 7 October 2022. Available at https://about.bgov.com/news/preventing-nuclear-armageddon-hinges-on-us-policy-of-ambiguity/
5. Budjeryn M (2022) Distressing a system in distress: global nuclear order and Russia's war against Ukraine. Bull At Sci 78(6):339–346
6. Savolainen R (2023) Seeking and sharing information about the threat of nuclear war. J Librariansh Inf Sci. https://doi.org/10.1177/09610006231219245
7. Kristensen HM, Korda M (2021) Russian nuclear weapons. Bull Atom Sci 77(2):90–108. https://doi.org/10.1080/00963402.2021.1885869
8. State Statistics Service of Ukraine (2024). https://ukrstat.gov.ua/
9. Zelensky names main task of audit of structural units of MoD and AFU. https://www.ukrinform.net/rubric-economy/3832172-zelensky-names-main-task-of-audit-of-structural-units-of-mod-and-afu.html
10. Muller J, Serruys P, Grubb N, Pastore J (2023) Prevention of nuclear war: a call to action for cardiovascular professionals. Eur Heart J 44(333–335). https://doi.org/10.1093/eurheartj/ehac702. https://academic.oup.com/eurheartj/article-pdf/44/5/333/49088413/ehac702.pdf
11. M Bivens (2022) Nuclear famine. Available at https://www.ippnw.org/wp-content/uploads/2022/08/Nuclear-Famine-2022.pdf

Crisis Communication Strategy in Case of a Nuclear Incident in Ukraine

Marin Gherman

Abstract The present article examines the main theoretical paradigms and applied strategies of crisis communication in the case of a nuclear incident, with a view to describing the specific situation in Ukraine in the context of the war against the Russian Federation. International experts have discriminated between communication in the pre-crisis phase, the crisis development *per se* as well as the post-incident phase, when the consequences of the nuclear incident are liquidated. It's worth mentioning that Ukraine has adopted a series of pre-emptive measures, based on a number of factors that favor an efficient communication strategy. Following the nuclear accident at Chernobyl, Ukraine preserved civil security classes in its education system. On the other hand, due to the war, communication interactions between the authorities and citizens have become more operative. At the same time, a number of obstacles might affect the efficiency of crisis communication, such as war fatigue, propaganda, social disappointment, waning trust in the authorities, Russia's destruction of energy and information infrastructure in Ukraine etc.

1 Introduction

Despite the countless studies and documents drafted at national or international levels, the effective management of crisis communication in the case of nuclear incidents poses a great social and political challenge. A key element is preparing the population in the pre-crisis phase in order to streamline communication strategies in the event of a nuclear accident. The successful implementation of measures designed to deal with contingencies depends on how well administrative authorities (governments in particular) communicate with the public (society) in emergency situations.

In this context, we should factor in both people's diverging perceptions of information delivered during the crisis, as well as the elaboration of communication strategies in the pre-crisis phase. A number of factors impact the effectiveness of

M. Gherman (✉)
Institute of Political Studies and Social Capital, Chernivtsi, Ukraine

F.-C. Cirstoiu et al., *Medical Response Strategy in Case of Radiation Emergency Caused by the War in Ukraine*, NATO Science for Peace and Security Series B: Physics and Biophysics, https://doi.org/10.1007/978-94-024-2266-5_10

crisis communication strategies, such as the government's ability to understand how people perceive official messages, as well as societal trust in the authorities (Perko et al. [16], p. 1998). Governments need to plan ahead strategic communication strategies in order to avoid widespread panic and to successfully coordinate all actions required to overcome the crisis. Recent studies show that communication planning should take under advisement the national specificity of each nation, while at the same time paying heed to international agreements, technical standards, ethical aspects as well as civil security regulations [17].

In that respect, the International Atomic Energy Agency (IAEA), as well as other researchers and experts at global level have published a number of recommendations concerning the way crisis communication should be managed. It's worth mentioning that most of the IAEA's recommendations refer to procedures in times of peace, namely to possible explosions at nuclear power plants or to unforeseen nuclear attacks. There are very few recommendations or studies that refer to nuclear incidents that might occur during large-scale military operations as part of prolonged wars like the one in Ukraine.

On 4 March 2022, the Russian army occupied the Zaporizhzhia Nuclear Power Plant (NPP), the biggest of its kind in Europe. Additionally, the Russian army also entered the territory of the Chernobyl nuclear power plant from the first day of the large-scale invasion of Ukraine. Following these two events, Ukraine's president, Volodymyr Zelenskyy, said that "no one can feel safe anywhere". The Kyiv leader called for a "global control" of Russia's nuclear installations and technologies in the context of the Russian Federation's repeated threats regarding the launch of a nuclear attack [19]. IAEA experts have visited the Zaporizhzhia Nuclear Power Plant on repeated occasions, examining various sites, including cooling stations, each time reporting they were denied full access to all facilities [22].

From this point of view, analyzing Ukraine's communication strategy in the event of a nuclear incident, caused either by an explosion at a nuclear power plant, or by the use of nuclear weapons, enables an estimate of future developments in the context of the war. In the following sections we will describe measures Ukraine has taken in this respect and analyze favorable and unfavorable factors at the level of Ukrainian society and the political class in order to identify the makings of an efficient crisis communication strategy.

2 Pre-emptive Communication Measures in the Case of a Nuclear Incident in Ukraine

In the textbook titled *Communication with the Public in a Nuclear or Radiological Emergency*, IAEA experts pointed out that crisis communication processes not only refer to the mere exchange of information, but rather to a complex causality that generates cognitive, emotional and behavioral disturbances. In the event of a crisis, IAEA recommends consulting all possible sources of information: radio and television, newspapers, social media, warning sirens, mobile speakers, hotlines set up to answer citizens' inquiries, public information centers, meetings and gatherings of various social groups, etc. [10].

At the same time, in another textbook, *Method for Developing a Communication Strategy and Plan for a Nuclear or Radiological Emergency*, IAEA experts recommend world states should elaborate a communication plan for emergency situations, which they deem necessary to streamline response actions in case of a nuclear incident. The plan ought to include a list of all the relevant institutions, of all stakeholders, while also designating a spokesperson that should liaise with the internal and external task forces, set up with the observance of specific procedures and regulations. IAEA provides member states with counseling regarding relations with mass-media, organizing press briefings or promoting various messages on social media [11].

Broadly speaking, the crisis communication strategy in case of a nuclear incident in Ukraine should comprise a number of key elements: communication in the pre-crisis phase (establishing the communication framework) in order to avoid panic; implementing the authorities' response algorithms with regard to the crisis; constant communication during the nuclear incident and its liquidation; consolidating and maintaining a relationship of trust between the authorities and society; external communication (Ukraine—international community); striking a balance between exhaustive / transparent communication and a limited communication in the context of the war. The communication strategy in the event of a nuclear incident should be all-encompassing, operative and straightforward, so as not to leave any room for interpretation.

It's worth dwelling on the pre-emptive character of crisis management. According to Nudell and Antokol [15], a number of pre-emptive measures are required to enable governments facing crises to coordinate their response. In other words, it's key to prepare the population beforehand for every possible outcome of a crisis. In this respect, an efficient crisis communication strategy depends on both the context before the crisis, as well as on how the public perceives official statements.

Starting the fall of 2021, Russia has been increasingly resorting to an anti-Western nuclear rhetoric. The closer the day of the large-scale invasion drew, the more Russia ramped up its rhetorical aggression. According to a thorough analysis of narratives Russia launched around the invasion and throughout the first year of the large-scale war, some 90 verbal interactions were reported between Russia and the West with respect to a possible nuclear conflict, aside from hundreds of other "relevant interactions" [1].

Ukrainian authorities carried out several specific activities in the pre-crisis phase. Their aim was to boost the resilience of society and avoid a potential society-wide panic. A number of government-led training and awareness-raising campaigns unfolded. After the launch of Russia's full-scale invasion of Ukraine, newspapers printed articles, while radio and TV stations broadcast segments meant to shore up national resistance. Some regional news agencies, for instance, wrote that the nuclear weapon "should not be considered a magical one, by means of which Russia will succeed in defeating Ukraine" [23]. The article in question examines a hypothetical nuclear strike launched by Russia against the town of Khmelnytskyi in Western Ukraine. According to the press article, despite the destruction, the Russian attack will do little to alter the course of the war, while Kyiv's defensive efforts will not be deterred.

The *Ukrainska Pravda* state-owned newspaper published a number of recommendations regarding civilian behavior in the event of a nuclear strike or explosion at a nuclear power plant. The article addresses teachers and headmasters in the secondary education cycle first and foremost, in order to inform them what they should do in case of a nuclear incident: how to protect pupils, where to hide, etc. The publication cites a specialist from the Ministry of Education in Kyiv, who urges participants in the educational process to consult the recommendations of the State Service for Emergency Situations. "An algorithm must be developed, so that in the event of an emergency evacuation, caused by nuclear incidents, we can ensure a safer connection between parents and children and establish a gathering place", the media article reads [3].

At the same time, the Ministry of Defense of Ukraine advised citizens not to panic when reading such information in the press. The 2021–2023 Ukrainian Minister of Defense, Oleksii Reznikov, said that the likelihood of the Russian Federation using nuclear weapons against Ukraine is low. That does not mean, however, that Ukraine should not be prepared. The government official added that the use of nuclear weapons by the Russian army "will open the ultimate frontline that will be the end of the aggressor state" [18].

With the aim of coordinating communication strategies in various crisis situations, the Center for Strategic Communication and Information Security, set up in Kyiv in 2021, developed the *Dovidka* platform. This online platform contains a series of information on recommended actions in the event of a nuclear disaster. According to the instructions of the government center, it is paramount citizens must closely follow the authorities' press statements on the radio, television or online platforms [7]. The strategic communication tools proposed by the Ukrainian government are accessible on various social networks and are not limited to the Internet. Radio and television stations broadcast official information posted on the *Dovidka* platform. This information is also disseminated on screens in trains, subways, buses and roadside screens in major cities. The platform developed by the Ukrainian government includes information on the location of shelters, the threat of nuclear incidents, cyber security and ways to identify disinformation and war propaganda.

In emergency situations, citizens are urged to keep calm. In the chapter titled "Connecting and searching for information", the Ukrainian government advises citizens to prepare for the scenario where the Internet and mobile connections become unavailable. "You need to install on your devices offline maps that will remain functional in the event of a network failure". "You should consider the possibility of purchasing a backup feature phone, which will work better than a smartphone", according to the platform developed by the Ukrainian authorities [6].

By systematically providing updated information to the public, the Ukrainian authorities gradually turned the topic of a nuclear incident into a method of encouraging resistance. According to a survey conducted in Ukraine at the start of 2023,

89% of Ukrainian respondents said that they would continue to fight against the aggression of the Russian Federation, despite a possible nuclear attack [4]. The authors of the study commissioned by the Munich Security Conference noted the resilience of society and "the unity of the Ukrainian population, based on the determination to protect their own democracy from Russia's autocratic revisionism".

It is also worth noting that the situation within Ukrainian society remains volatile and can change from one day to the next. The longer military actions last, the more relevant a number of new factors will become, such as war fatigue or disinformation. However, Ukraine is an example of how a security threat (the nuclear strike, the danger of an explosion at a nuclear power plant) can serve to encourage popular resistance with the help of crisis communication strategies.

3 Factors Favoring an Efficient Crisis Communication Strategy

The war in Ukraine is in itself a favorable factor for social resilience in the face of a nuclear incident, despite the fact that military operations represent an atypical context for an efficient communication. The annexation of the Crimean Peninsula by the Russian Federation in 2014 and the 2014–2022 war in Eastern Ukraine have accustomed the population to the feeling of insecurity. Moreover, after Russia launched the full-scale invasion of Ukraine on February 24, 2022, the population has become more resistant to military or non-military threats. Ukrainian psychologists describe the process whereby civilians get used to negative war-related news as "emotional numbing" [12]. This social phenomenon somewhat reduces the shockwave compared to periods of stability and tranquility.

On the other hand, "emotional numbing" alters the perception of Ukrainian society with regard to media reports of casualties or potential threats. According to Kyiv-based journalists who covered the events of the war *in extenso*, starting 2022 Ukrainian society has gradually got used to negative news. According to Anton Semyzhenko, the coordinator of the English-language department of the Kyiv-based publication *Babel*, journalists no longer highlight in bold the newsfeed with reports of two or three people killed by the Russian army in the context of military operations. "Such information would have had a completely different impact in times of peace", Semyzhenko argues, also pointing to a change in the perception of Ukrainian society towards social and political realities [8].

Another factor conducive to an efficient crisis communication strategy in the event of a nuclear incident has to do with the positive experience of authorities in overcoming certain drawbacks in the context of war. We have witnessed a number of major challenges to the security of the state and various difficult situations. The positive phases of the war for Ukraine include the following: repelling the invasion of the Kyiv region in the spring of 2022, preventing the occupation of Kharkiv, repelling intensive artillery attacks in Donbas, the successful counteroffensive in the south and east of Ukraine, providing assistance to people affected by the destruction of the Nova Kakhovka dam. Ukrainian authorities have been able to overcome these challenges and others with the help of warfare-specific communication techniques.

During this period, social media posts were very important, reflecting the multi-layered relation between people and state authorities. Social media has given people a chance to rebuild their relationship with the state online, based on the common challenges they face. A number of posts show how people support each other, for example by creating invincibility centers, hosting fundraisers or taking other specific actions [9]. Both Ukrainian government institutions and non-governmental organizations or Ukrainian individuals posted messages on social media, fostering the image of heroes who defend their country or give their lives to ensure the survival of their nation. Their posts often included pictures, lyrics, metaphors, quotes, jokes and other strategies designed to bolster resistance.

At the same time, it is worth discussing how Ukrainian society experienced the Chernobyl nuclear accident. After the 1986 tragedy, the USSR introduced civil security classes in the secondary and higher education systems. These classes were carried over and developed by the Ukrainian education system. Starting 1991, students in Ukraine have been told what to do in the event of a nuclear attack, how to react to such a crisis, how to interact with the military, where to hide if a nuclear power plant explodes or a nuclear bomb is launched. All this information could help avoid a mass panic in the event of a crisis. For example, the Civil Security textbook, which was used in class by all students of the Lviv Polytechnic National University in 2010, long before Russia's full-scale invasion of Ukraine, contains a range of information on nuclear incidents. Whole chapters of the manual are devoted to the analysis of radioactive leaks, the liquidation of the nuclear fallout of the accident and communication with the authorities or the military [21].

Textbooks addressing high school students in Ukraine also contain a lot of information on civilian protection procedures in the event of a nuclear incident. In the Homeland Defense textbook [2] we also find information about evacuation procedures in the event of a nuclear incident, the need to avoid panic in crisis situations, but also about the state's communication channels. School curricula in Ukraine include information on the first measures citizens need to take in the event nuclear sirens go off. "You must turn on the radio or television in order to receive updated emergency notifications from the authorities", the textbooks reads.

The Chernobyl nuclear accident itself and the liquidation of its fallout starting 1986 have made Ukrainians weary of a potential reoccurrence of such a scenario, as sociological surveys also show [4]. The retention of curricula in Ukraine's education system on civilian protection measures following a nuclear incident, as well as the relatively recent trauma of the Chernobyl nuclear accident make Ukrainian society less prone to panic.

4 Possible Obstacles to Efficient Crisis Communication Management

Whereas factors that favor an efficient crisis communication strategy in the event of a nuclear incident are specific to the Ukrainian state or society, at the opposite end we have those factors that are directly linked with the duration and nature of the war. One of the most important obstacles to efficient crisis communication is war fatigue. The military conflict is, on the one hand, a favorable factor, and on the other hand, one that could diminish vigilance and social resilience the longer hostilities last.

According to a survey conducted by the Kyiv International Institute of Sociology in the fall of 2023, Ukrainians seem much more exposed to the propaganda of the Russian Federation due to war fatigue. In the 2022–2023 period, the number of Ukrainians who are affected by various false narratives promoted by the Russian Federation has increased. More and more civilians believe that Western states have grown weary of Ukraine and are asking the authorities in Kyiv to come to an understanding with Russia. At the same time, the study shows that the number of people who fear a betrayal of the authorities, in the sense of Kyiv accepting undesired compromises or especially territorial concessions, has increased [14]. According to Ukrainian sociologists, Russian propaganda became more effective as the war dragged on, impacting the population's trust in the authorities.

War fatigue as a social and political factor was also recognized by representatives of the Ukrainian government. In September 2023, Oleksiy Danilov, Secretary of the Security and Defense Council of Ukraine, compared the war in Ukraine to sports, noting the fatigue that builds up [5]. "We are now at a strategic turning point. Everyone expected a sprint, but the war is turning into a marathon, requiring the athlete to meet certain fitness requirements and undergo a different training regimen", the Kyiv official wrote in an article published in the national newspaper *Ukrainska Pravda*. Danilov cited a series of false narratives promoted by the Kremlin to confuse Ukrainians and generate mistrust in the authorities.

In this context, it is worth mentioning that Russia's fake news, propaganda and disinformation, directed against Ukraine in the context of the war, represent a significant obstacle in the way of an efficient management of crisis communication. Combined with the dissemination of fake news, war fatigue has led to an increase in the level of distrust in the authorities in the second year of the military conflict. Surveys show that in 2023, Ukrainians started being more critical and distrustful of the political class, while at the same time society preserved a maximum level of trust in the military. Confidence in the Armed Forces of Ukraine broke all records in the history of the Ukrainian state, reaching 96% [13]. Surveys carried out on the territory of Ukraine show that society is more willing to collaborate with the representatives of the army, rather than with the political class.

Another factor that must be taken under advisement when elaborating the crisis communication strategy in the event of a nuclear incident is the possible destruction of the information infrastructure. Traditional communication channels might get blocked or jammed. Therefore, authorities need to take into account all the relevant recommendations of international organizations, especially the IAEA. In the context of the war in Ukraine, there have been reports of citizens being completely cut off from the authorities following power outages and the discontinuity of Internet and mobile phone networks.

Testimonies published by Ukrainian national media are telling in this regard. According to the *Obozrevatel* newspaper, after residents of the Bucha district in the Kyiv region were left without water, electricity and Internet on February 25, 2022, people were disconnected from any forms of communication with the Ukrainian authorities. "I found an old battery-powered radio in the attic and I listened to our stations for a while. The authorities didn't say anything about an evacuation, that's why we didn't leave the house for a while", according to the testimony of a Ukrainian woman [20]. Thanks to traditional means of communication (radio, television), some civilians survived the military occupation. Therefore, in times of crisis, the authorities should resort to all means of communication, despite the fact that some may be considered technologically obsolete or futile.

5 Conclusions

Ukraine adopted a series of pre-emptive measures in the context of the war with Russia, preparing its population for a possible nuclear incident. With the help of wartime communication methods, the Ukrainian authorities managed to transform nuclear threats into resilience-building factors at society level.

Factors conducive to an efficient crisis management strategy in case of a nuclear incident include the following: the experience gained during the war, the population growing accustomed to the state of instability, preserving civil security classes in public schools, and the recent trauma caused by the explosion at the nuclear power plant in Chernobyl. Factors that hinder an efficient communication strategy in case of a nuclear incident include: growing war fatigue, disinformation and propaganda, waning trust in the authorities and technical difficulties associated with emergency situations.

Despite possible communication barriers, social resilience in Ukraine with regard to a nuclear scenario appears to be quite high. At the same time, against the backdrop of waning trust in politicians and hybrid attacks by means of propaganda and disinformation, Ukrainian society fully confides in the military. Namely, the military, not the political class, is directly responsible for managing crisis communication during a potential nuclear incident. In this sense, Ukraine is a good example of how a security threat (a nuclear strike, the threat of an explosion at a nuclear power plant) can serve to encourage popular resistance with the help of crisis communication strategies.

References

1. Arndt A, Horovitz L (2022) Nuclear rhetoric and escalation management in Russia's war against Ukraine: a chronology. SWP Working Papers 03:1–100
2. Baka M, Kvashnov Y, Lytvynenko A (2006) Zakhyst Vitchyzny. Vesna, Kyiv
3. Barsukova O (2022) Yak pidhotuvatysia do yadernoi zahrozy ta evakuiuvaty ditei? Instruktsiia osvitnoho ombudsmena, Ukrainska Pravda. https://life.pravda.com.ua/health/2022/10/5/250708
4. Bunde T, Eisentraut S, Schütte L (2023) Auch bei russischem Nuklearschlag: 89 Prozent der Ukrainer wollen unter allen Umständen weiterkämpfen. Tagesspiegel. https://www.tagesspiegel.de/internationales/auch-bei-russischem-nuklearschlag-89-prozent-der-ukrainer-wollen-weiterkampfen%2D%2Dbis-zur-ruckeroberung-der-krim-9299993.html
5. Danilov O (2023) Stratehichna tochka perelomu. Ukrainska Pravda. https://www.pravda.com.ua/columns/2023/09/16/7420134
6. Dovidnyk (2022a) Susidski chaty: yak dilytys informatsiieiu bezpechno. Dovidka. https://dovidka.info/yak-diyaty-v-razi-zastosuvannya-brudnoyi-bomby-yadernoyi-ataky-chy-avariyi-na-aes
7. Dovidnyk (2022b) Yak vriatuvatys vid brudnoi bomby, zastosuvannia yadernoi zbroi chy avarii na AES. Dovidka. https://dovidka.info/yak-diyaty-v-razi-zastosuvannya-brudnoyi-bomby-yadernoyi-ataky-chy-avariyi-na-aes
8. Gherman M (2022) Jurnalist din Kiev, despre lupta de pe frontul presei din Ucraina: „Am lucrat non-stop, fără libere, fără salarii. Viața oamenilor depinde de o informare corectă". Libertatea. https://www.libertatea.ro/stiri/jurnalist-din-kiev-despre-lupta-de-pe-frontul-presei-din-ucraina-am-lucrat-non-stop-fara-libere-fara-salarii-viata-oamenilor-depinde-de-o-informare-corecta-4242202
9. Goodman B (2023) Russian warship, go F*ck yourself. Circulating social media discourses in the Russia–Ukraine War. Dispossession Anthropological Perspectives on Russia's War Against Ukraine. Routledge https://doi.org/10.4324/9781003382607
10. IAEA (2012) Communication with the public in a nuclear or radiological emergency. EPR-Public Communications. https://www-pub.iaea.org/MTCD/Publications/PDF/EPR-Communcation_web.pdf
11. IAEA (2015) Method for developing a communication strategy and plan for a nuclear or radiological emergency. EPR-Public Communications. https://www-pub.iaea.org/MTCD/Publications/PDF/EPR-CommPlan2015_web.pdf
12. Karbunar N (2022) Emotsiine oniminnia. Psykholoh rozpoviv, shcho naspravdi kryietsia za zvykanniam do viiny. Glavkom. https://glavcom.ua/country/society/emotsijne-oniminnja-psikholoh-rozpoviv-shcho-naspravdi-krijetsja-za-zvikannjam-do-vijni-869917.html
13. KIIS (2023a) Dynamika doviry sotsialnym instytutsiiam u 2021–2023 rokakh. KIIS. https://www.kiis.com.ua/?lang=ukr&cat=reports&id=1335
14. KIIS (2023b) Dynamika poshyrenosti sered naselennia naratyviv rozkolu: rezultaty telefonnoho opytuvannia, provedenoho 29 veresnia—9 zhovtnia 2023 roku. KIIS. https://www.kiis.com.ua/?lang=ukr&cat=reports&id=1302
15. Nudell M, Antokol N (1988) The handbook for effective emergency and crisis management. Lexington Books
16. Perko T, van Gorp B, Turcanu C, Thijssen P, Carle B (2013) Communication in nuclear emergency preparedness: a closer look at information reception. Risk Anal 33(11): 1987–2001. https://doi.org/10.1111/risa.12048
17. Razucka F, Escarania P, dos Santosa R (2023) The strategic communication planning in an emergency at the Almirante Álvaro Alberto nuclear power plant (CNAAA). Braz J Radiat Sci 11(1):1–17. https://doi.org/10.15392/2319-0612.2023.2148
18. SD (2022) Reznikov zaklykav svitovi krainy prypynyty boiatysia druhu armiiu svitu. Slovo i dilo. https://www.slovoidilo.ua/2022/10/06/novyna/polityka/reznikov-zaklykav-svitovi-krayiny-prypynyty-boyatysya-druhu-armiyu-svitu

19. Stanciu E (2022) Zelenski: După tot ce a făcut armata rusă la Cernobîl şi la centrala nucleară Zaporojie, nimeni din lume nu se poate simţi în siguranţă. Libertatea. https://www.libertatea.ro/stiri/zelenski-dupa-tot-ce-a-facut-armata-rusa-la-cernobil-si-la-centrala-nucleara-zaporojie-nimeni-din-lume-nu-se-poate-simti-in-siguranta-4100972
20. Starokon S (2022) Dity pislia zvilnennia tsiluvaly khlib, a rashysty vbyvaly navit sobak: ukrainka pro okupatsiiu Kyivshchyny. Obozrevatel. https://war.obozrevatel.com/ukr/diti-pislya-zvilnennya-tsiluvali-hlib-a-rashisti-vbivali-navit-sobak-ukrainka-pro-okupatsiyu-kiivschini.htm
21. Vasiichuk V, Honcharuk V, Kachan V, Mokhniak S (2010) Osnovy tsyvilnoho zakhystu. Publishing House of the Lviv Polytechnic National University
22. Vidu V (2023) Grossi, în vizită la centrala Nucleară Zaporojie pentru a evalua situaţia în urma distrugerii Barajului Kahovka. Newsro. https://www.news.ro/externe/grossi-in-vizita-la-centrala-nucleara-zaporojie-pentru-a-evalua-situatia-in-urma-distrugerii-barajului-kahovka-1922402315002023061621228143
23. Yashchyshen A (2022) Yak vriatuvatysia vid yadernoho vybukhu. YE.ua. https://ye.ua/syspilstvo/60999_Yak_vryatuvatisya_vid_yadernogo_vibuhu.html

Part II
Shelling Ukraine's Nuclear Power Plant, Radioactive Waste Storage, Hospitals, Research Institutes or Factories That Use Radioactive Sources. Reaching Medical Preparedness for a Radiation Emergency

Nuclear and Radiological Events: Should We Consider Individual Factors?

Victor Trigon, Michel Bourguignon, and Nicolas Foray

Abstract Our recent radiobiological study with a cohort of apparently healthy individuals reveals that more than 20% of them show an abnormal response to ionizing radiation (IR) leading to high risk of radiation-induced (RI) adverse tissue reactions and/or cancer. These data are supported by a mechanistic model based on the (RI) nucleoshuttling of the ATM protein (RIANS) that provides a relevant biological interpretation of the individual response to IR. Such findings are also consistent with former estimations of a sub-population of radiosensitive or radiosusceptible adults representing 5–20% of the whole population. In parallel, international guidelines of radiological protection still consider individuals as uniformly radioresistant: current biological dosimetry process applied after nuclear and radiological events may seriously underestimate the risks of exposure to IR of such sub-populations. Here, we discussed on the possibility of a rapid detection and specific management of abnormally responding individuals in a mass crisis context of a nuclear and radiological event. Ethical issues will be also addressed.

1 Introduction

Our recent clinical investigations on skin fibroblasts from 30 apparently healthy male volunteers of the staff of an emergency medical unit of Paris, France, revealed a subset of 7 (> 23%) individuals whose molecular and cellular response to ionizing radiation (IR) appeared abnormal i.e. suggests high risk of radiation-induced (RI) toxicity and cancer (Foray et al. in preparation). These data are in agreement with a long series of studies suggesting, notably in the context of medical exposures, that a non-negligible (5–20%) sub-population of radiosensitive adults does exist (Table 1).

V. Trigon · M. Bourguignon · N. Foray (✉)
U1296 Unit "Radiation: Defense, Health and Environment", Institute of Health and Medical Research (INSERM), Centre Léon-Bérard, Lyon, France
e-mail: nicolas.foray@inserm.fr

F.-C. Cirstoiu et al., *Medical Response Strategy in Case of Radiation Emergency Caused by the War in Ukraine*, NATO Science for Peace and Security Series B: Physics and Biophysics, https://doi.org/10.1007/978-94-024-2266-5_11

Table 1 Estimations of the sub-populations of radiosensitive individuals (in percentages)

References	Group I	Group II	Group III
[4]	77 ± 3	17 ± 3	0.8 ± 0.1
[30]	71.6	23.9	4.5
[2]	82	17.5	0.5
Foray et al., in preparation	76.7	23.3	0
Average	76.8 ± 4	20.4 ± 2	1.4 ± 1

Furthermore, such statement is also consistent with a mechanistic model based on the (RI) nucleoshuttling of the ATM protein (RIANS), a major kinase required for the molecular and cellular response to genotoxic stress. Since international guidelines of radiological protection still consider individuals as uniformly radioresistant [6-8], the current protocols established for managing irradiated individuals after a nuclear or radiological event may seriously underestimate such a sub-population at risk. Here, we discussed whether rapid and specific detection and management of these abnormally IR-responding individuals are possible in a mass crisis context of a nuclear or radiological event.

2 The Major Clinical Consequences of an Exposure to Ionizing Radiation

After more than one century of data acquisition, 3 major clinical consequences of an exposure to IR can be defined: radiosensitivity, radiosusceptibility and radiodegeneration:

3 Radiosensitivity Reactions (or Radiotoxicity)

Definition: The reactions of **radiosensitivity** (often called radiotoxicity in the context of nuclear or radiological events) are defined at the tissue scale as the adverse tissue effects (burns, dermatitis, proctitis, . . .) that can lead to a dysfunction of the irradiated tissue/organs, and, in the most extreme cases, to the death of the irradiated individual. ***Causes:*** Radiosensitivity is generally attributable to cell death at the cellular and tissue scale, to chromosome breaks at the cytogenetic scale and to unrecognized and/or unrepaired DNA double-strand breaks (DSB) at the molecular scale. ***Conditions of occurrence:*** Radiotoxicity reactions require *both* doses higher than 0.7 Gy and dose-rates higher than 0.1 Gy/min to occur (i.e. of the same order as the most current radiotherapy (RT) modalities, or, in more extreme conditions, like nuclear power plant accident or atomic bomb explosion). For doses higher than 4 Gy delivered to the entire body, radiopathology describes the well-documented acute radiation syndrome (ARS). ARS is sub-divided in three steps: the prodromal, latency and critical periods. During the critical period, bone marrow injury and changes in the blood composition start at 0.7 Gy, gastrointestinal syndrome occurs above 6 Gy and neurovascular injury occurs for doses greater than 8 Gy. Doses to the entire body higher than 10 Gy are immediately lethal [10].

4 Radiosusceptibility (or Radiation-Induced Cancers)

Definition: The first RI cancers have been described very early after the X-rays discovery, suggesting that the pioneers of radiation were aware of the different natures of the RI risk (toxicity vs cancer) [11]. ***Causes:*** **Radiosusceptibility**, defined as the RI cancer proneness, is generally attributable to cellular transformation facilitated by both amplification of gene mutations and impairment of cell cycle checkpoint arrests. At the cytogenetic and molecular scales, radiosusceptibility is attributable to chromosome aberrations and misrepaired DSB, respectively. ***Conditions of occurrence:*** Occurrence of RI cancer as a function of dose is at the centre of the controversies between the linear-non threshold (LNT) and non-linear with threshold (NLT) models [12–17].

From Hiroshima survivors data, when a single dose is applied in a very short time (some seconds), the risk of RI leukemia and solid cancers is known to increase from a threshold dose of 100 and 200 mSv, respectively. In the medical context, RI cancers may be observed in surviving tissues either after RT, after repeated CT scan (a few dozen mGy per examination) or mammography (2 mGy per view) exams. RI cancers can also be observed after sublethal doses after accidental or occupational exposure.

5 Radiodegeneration (or Radiation-induced accelerated aging)

Definition: **Radiodegeneration** has been defined as the proneness to RI accelerated aging of specific tissues. ***Causes:*** Radiodegeneration is generally attributable to a senescence process at the cellular scale and to tolerated but unrecognized DNA strand breaks that accumulated in the cell nucleus. The most documented radiodegeneration reactions are RI cataracts but some RI aging features have been already observed in bone and cardiovascular system. ***Conditions of occurrence:*** like for RI cancers, the RI accelerated aging features occur at sublethal doses. However, the process of aging is particularly favoured when exposure to IR is extended in time (i.e. at low dose-rates). As an example, RI cataracts observed in astronauts are the most probable consequence of a continuous exposure to space radiation. Forty-eight cases of severe lens opacification (16.2%) were observed among the 295 NASA astronauts who participated to the LSAH (Longitudinal Study of Astronaut Health) study, but 86% of the astronauts who stayed in space suffer from a pathology of the eye. An increased risk of cataract was generally observed after exposure to doses higher than 8 mGy, which corresponds to approximately 20 mission days [19–22].

6 Predisposing Factors and Nuclear and Radiological Incidents

As evoked above, radiosensitivity, radiosusceptibility and radiodegeneration occur in specific dose- and dose-rate ranges but also share 2 common features [1]: the molecular biomarkers that would be specific of each of the 3 evoked notions are either undefined or not consensual yet, and the individual predisposition for each of the 3 notions, notably the potential genetic predisposing factors remain to be clarified further or are still to be determined. In addition, it appears obvious that accelerated aging is likely to be observed for protracted, continuous or chronic exposures to IR, which may be the case for post-accident contaminated areas but not necessarily for a short-term management of nuclear or radiological events delivering high doses. Furthermore, the follow-up of the patients suffering from RI cataracts, even predisposed genetically, appears to be less critical than for RI toxicity and cancer. Hence, for the rest of this review, we will focus on individual radiosensitivity and radiosusceptibility.

7 Quantitative Features of Individual Radiosensitivity and Radiosusceptibility

7.1 *Individual Radiosensitivity*

Most of knowledge about individual radiosensitivity has been provided from post-RT data. Since some decades, consensual severity grades scale have been defined to quantify the severity of the tissue radiosensitivity observed on healthy tissues after RT. One of the most current scale, the common terminology criteria for adverse effect (CTCAE) scale permits to score post-RT radiotoxicity reactions from grade 0 (no reaction) to grade 5 (death of patient) [25]. The considerable advances in RT technologies have not permitted to reduce drastically the proportion of grade 3–4 reactions that still represents a significant subset of clinical radiosensitivity reactions estimated to about 5–20% cases, whatever the RT modality and the type of the tumour (Table 1).

Since the 1980s, the *in vitro* assessment of the clonogenic cell surviving fraction at 2 Gy (SF2) has been shown to predict the clinical radiosensitivity [26]. In humans, the SF2 range from 1%, the highest radiosensitivity, corresponding to the ataxia telangiectasia syndrome, to 60%, the normal response considered as radioresistance

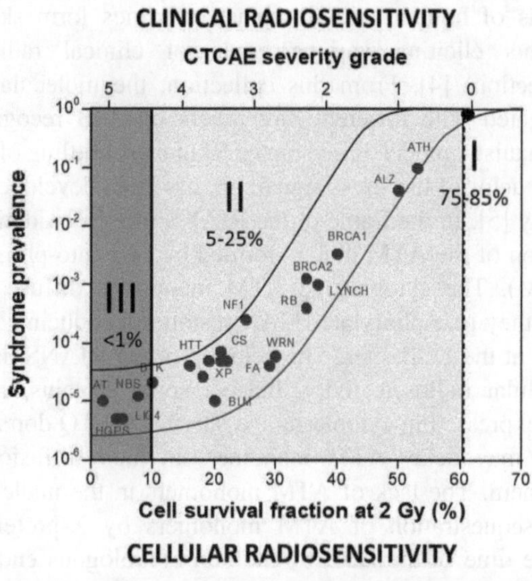

Fig. 1 Syndrome prevalence as a function of SF2. All the SF2 values were provided from and the corresponding prevalence values are available in literature (epidemiological data). For convenience, the error bars were omitted. The corresponding CTCAE grades are also indicated by using formula linking SF2 and grades. The grey and white zones indicate the 3 groups of radiosensitivity defined by SF2 values ranging in 0–10%, 10–60% and > 60% survival, respectively. The abbreviations of syndromes are: *AT,* ataxia telangiectasia, *LIG4,* LIG4 syndrome, *HGPS,* Hutchinson-Gilford progeroid syndrome, *NBS,* Nijmegen's syndrome, *BTK,* Bruton kinase syndrome, *BLM*, Bloom's syndrome, *HTT,* Huntington's disease, *XP,* xeroderma pigmentosum syndromes, *CS,* Cockayne's syndromes, *FA,* Fanconi anemia, *WRN,* Werner's syndrome, *NF1,* neurofibromatosis type 1, *RB,* retinoblastoma syndrome, *DOWN,* Down's syndrome, *LFS,* Li-Fraumeni's syndrome, *BRCA2,* *BRCA2* mutations, *BRCA1,* *BRCA1* mutations, *LYNCH,* Lynch's syndrome, *ALZ,* Alzheimer's disease, *ATH,* heterozygous mutations of the *ATM* gene

[24]. If SF2 is plotted against the prevalence of the genetic syndromes, associated with different levels of radiosensitivity, 3 groups of cellular radiosensitivity appear (Fig. 1):

- one group corresponding to SF2 > 60%, representing 75–85% patients, considered as radioresistant, i.e. at low risk of radiotoxicity reactions;
- one group gathering syndromes associated with moderate radiosensitivity with 10% < SF2 < 60% representing 15–20% individuals, and generally caused by heterozygous genetic mutations leading to an over-expression of the mutated protein and its abnormal sub-cellular localization;
- one group gathering syndromes associated with hyper-radiosensitivity (SF2 < 1%) that represent less than 1% individuals, and generally caused by homozygous genetic mutations leading to a definitive loss of biological function.

However, the clonogenic cell survival assay is too-time consuming to be applied in routine as a predictive assay. That's why radiobiologists have focused their efforts on molecular assays to predict the individual radioresponse. For 20 years, we have collected hundreds of human skin fibroblast cell lines form skin biopsies from RT-treated patients eliciting various grade of clinical radiosensitivity (the COPERNIC collection) [4]. From this collection, the molecular radiosensitivity has been investigated with different biomarkers of DSB recognition and repair pathways. A mechanistic model, based on the RI nucleoshuttling of the ATM protein (RIANS), a major actor of the stress signalling, has been developed to predict post-RT radiosensitivity [5]. In the frame of the RIANS model, oxidative stress leads to the monomerization of the ATM dimers formed by two auto-phosphorylated ATM monomers (pATM). The cytoplasmic ATM monomers diffuse thereafter to the nucleus in which they phosphorylate H2AX histones (producing γH2AX immunofluorescence foci) at the DSB sites. The velocity of the RIANS has been shown to condition the cellular radiosensitivity. Indeed, some proteins, called X-proteins, abnormally over-expressed in cytoplasm and sharing SQ/TQ domains of phosphorylation by ATM may delay ATM monomers in their diffusion to nucleus by interacting with them. The lack of ATM monomers in the nucleus resulting from the cytoplasmic sequestration of ATM monomers by X-proteins, prevents the recognition in due time of the DSB by the non-homologous end-joining (NHEJ), the most predominant DSB recognition and repair pathways in humans. Three groups of molecular radioresponses have been defined (Fig. 2):

- the group I gathers radioresistant individuals whom fibroblasts elicit a rapid RIANS (some min after 2 Gy X-rays). The maximal number of pATM foci (pATMmax) is 40 ± 5 after 2 Gy X-rays followed by 10 min-1 h;
- the group II gathers individuals with moderate radiosensitivity whom fibroblasts elicit a RIANS delayed by the over-expressed X-proteins: $5 <$ pATMmax<35 foci by 10 min-1 h;
- the group III gathers hyper-radiosensitive individuals (mostly pediatrics syndromes) whom fibroblasts elicit gross defect in the ATM nucleoshuttling (pATMmax <5 by 10 min-1 h) or in DNA repair.

Our Research Unit has already classified several genetic syndromes associated with radiosensitivity and or radiosusceptibility with RIANS biomarkers [27–29]. Furthermore, the RIANS has permitted to predict post-RT radiosensitivity with a high statistical robustness [2, 4]. Interestingly, the COPERNIC collection and the RIANS model have also helped in estimating the relative proportion of each group of radioresponse: by investigating the different quantitative constraints of DSB recognition and repair biomarker values of human cells, the groups, I, II and III defined with the molecular criteria described above were found to represent 71.6 ± 4, 23.9 ± 4 and $1.5 \pm 1\%$ of the whole populations [3, 30]. More recently, a mathematical link has been established between SF2 and the RIANS values [2]. The definitions of the 3 groups of radioresponses at the cellular and at the molecular scale can therefore be merged (Table 1).

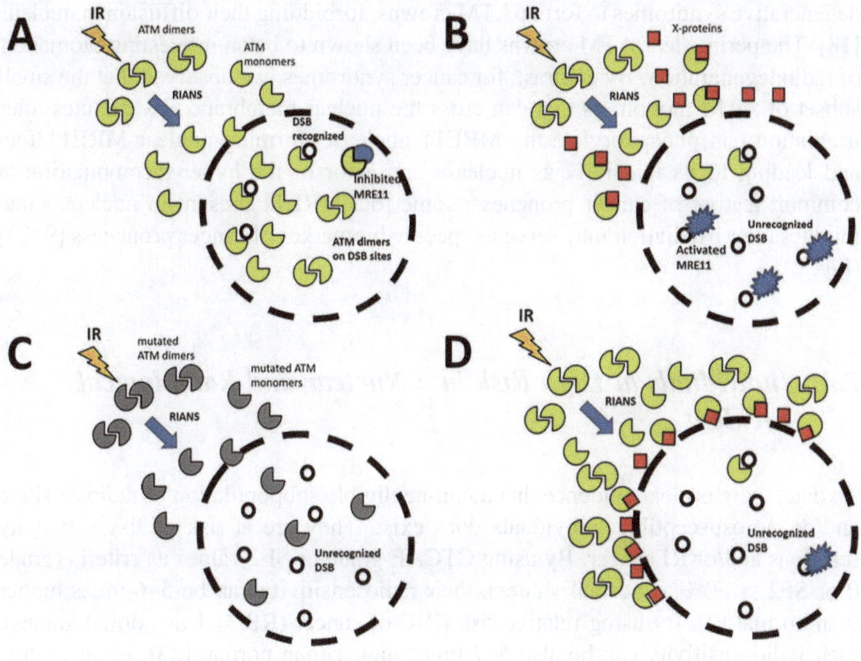

Fig. 2 Schematic illustration of the 3 groups of radiosensitivity defined by the RIANS model: (**a**) Group I: The RIANS is rapid and the DSB recognition is complete. Cells are radioresistant. (**b**) Group II with cancer proneness. The RIANS is delayed by the sequestration of ATM monomers by X-proteins in cytoplasm. MRE11 is activated and increases the number of misrepaired DSB. (**c**) Group IIIa. In the case of *ATM* mutations, the RIANS is not functional and no DSB are recognized. (**d**) Group II with aging: If X-proteins specifically bind to the nuclear membranes, the ATM-X complexes prevent any diffusion of ATM monomers and facilitate the re-dimerization of the ATM monomers around the nucleus. As a result, very few DSB are recognized and DSB cumulate to favour senescence death [9]. The panel B illustrates the opposite situation in which X-proteins are dispatched in cytoplasm

7.2 *Individual Radiosusceptibility*

As shown above, the delay in RIANS (group II) or its absence (group III) may lead to (hyper)radiosensitivity and cancer proneness or else (hyper-)radiosensitivity and degenerative diseases. For each group II and III, degenerative diseases represent a minority of syndromes by comparison to cancer syndromes. Hence, by considering error bars, the estimations shown in Table 1 remain also relevant for cancer syndromes alone. In the frame of the RIANS model, we have provided evidence that X-proteins of degenerative syndromes are specifically localised around nucleus (perinuclear localization) while X-proteins of cancer syndromes are rather dispatched in cytoplasm [9] (Fig. 2). Hence, early after irradiation, a subset of ATM monomers can diffuse in the nucleus in the case of cancer syndromes while ATM monomers agglutinate with X-protein all around the nucleus in the case of degenerative syndromes to form pATM crowns, forbidding their diffusion in nucleus [18]. The perinuclear ATM crowns have been shown to be an interesting biomarker of radiodegeneration. By contrast, for cancer syndromes, we observed that the small subset of ATM monomers that can cross the nuclear membrane few minutes after irradiation can phosphorylate the MRE11 nuclease, forming nuclear MRE11 foci and leading to its inactivity as nuclease and actor of the hyper-recombination, a common feature of cancer proneness: some foci MRE11 present in nucleus some minutes after irradiation may serve as specific biomarker of cancer proneness [9, 23] (Fig. 2).

7.3 Individuals at High Risk in a Nuclear and Radiological Incident

To date, there is clear evidence that a non-negligible subpopulation of radiosensitive and/or radiosusceptible individuals does exist. They are at risk of tissue toxicity reactions and/or RI cancer. By using CTCAE grade or SF2 values as criteria (grade 0 or SF2 $>$ 60% as normal values), their radiosensitivity can be 5–6 times higher than normal [2]. By using relative risk (RR) of cancer (RR $=$ 1 as normal values), their radiosensitivity can be also 5–7 times higher than normal [23]. Hence, while radiation oncologists take into account such clinical features to eventually modify the anti-cancer treatment planning, such subpopulation involved in nuclear or radiological events may raise practical and ethical problems with a specific and more complex follow-up. The International guidelines of radioprotection still considering individuals as equally radioresistant and at low risk of cancer will underestimate such a non-negligible subpopulation. Hence, significant efforts must be provided to detect rapidly and reliably the radiosensitive and radiosusceptible individuals during the management and the follow-up of victims exposed to IR for nuclear or radiological events. Here, we proposed first preliminary solutions of biomarkers specific to RI toxicity and cancer proneness and based on the RIANS model: pATMmax for toxicity and MRE11 foci for cancer proneness. However, these biomarkers are observed during immunofluorescence of established cell lines in culture, a technique which can be too time-consuming after a nuclear and radiological event. Biochemical procedures (e.g. ELISA test) may permit to alleviate the problem of cell culture and establishment of cell line by replacing these steps by cellular extracts. Further investigations are therefore needed to enable the rapid management of significant subpopulation of radiosensitive/radiosusceptible individuals after a nuclear or a radiological event.

References

1. Foray N, Bourguignon M, Hamada N (2016) Individual response to ionizing radiation. Mutat Res Rev
2. Le Reun E, Bodgi L, Granzotto A, Sonzogni L, Ferlazzo ML, Al-Choboq J et al (2022) Quantitative correlations between radiosensitivity biomarkers show that the ATM protein kinase is strongly involved in the radiotoxicities observed after radiotherapy. Int J Mol Sci
3. Ferlazzo M, Foray N (2024) Effets biologiques des radiations ionisantes. Tome 3 Maladies génétiques associées à la radiosensibilité. Editions universitaires européennes, Sarrebruck
4. Granzotto A, Benadjaoud MA, Vogin G, Devic C, Ferlazzo ML, Bodgi L et al (2016) Influence of nucleoshuttling of the ATM protein in the healthy tissues response to radiation therapy: toward a molecular classification of human radiosensitivity. Int J Radiat Oncol Biol Phys
5. Berthel E, Foray N, Ferlazzo ML (2019) The nucleoshuttling of the ATM protein: a unified model to describe the individual response to high- and low-dose of radiation? Cancers
6. ICRP (1999) Genetic susceptibility to cancer. Publication 79. Ann ICRP
7. ICRP (2007) Radiation protection in medicine. Publication 105. Ann ICRP
8. ICRP (2012) ICRP statement on tissue reactions and early and late effects of radiation in normal tissues and organs-threshold doses for tissue reactions in a radiation protection context. Publication 118. Ann ICRP
9. El-Nachef L, Al-Choboq J, Restier-Verlet J, Granzotto A, Berthel E, Sonzogni L et al (2021) Human radiosensitivity and radiosusceptibility: what are the differences? Int J Mol Sci
10. Dainiak N, Gent RN, Carr Z, Schneider R, Bader J, Buglova E et al (2011) Literature review and global consensus on management of acute radiation syndrome affecting nonhematopoietic organ systems. Disaster Med Public Health Prep
11. Frieben A (1902) Cancroid des rechten Handrückens. Deutsche medicinische Wochenschrift
12. Boice JD Jr (2017) The linear nonthreshold (LNT) model as used in radiation protection: an NCRP update. Int J Radiat Biol
13. Calabrese EJ (2013) Origin of the linearity no threshold (LNT) dose-response concept. Arch Toxicol
14. Preston DL, Shimizu Y, Pierce DA, Suyama A, Mabuchi K (2003) Studies of mortality of atomic bomb survivors. Report 13: solid cancer and noncancer disease mortality: 1950–1997. Radiat Res:2012
15. Shah DJ, Sachs RK, Wilson DJ (2012) Radiation-induced cancer: a modern view. Br J Radiol 85(1020):e1166–e1173
16. Hatch M, Cardis E (2017) Somatic health effects of Chernobyl: 30 years on. Eur J Epidemiol
17. Mathews JD, Forsythe AV, Brady Z, Butler MW, Goergen SK, Byrnes GB et al (2013) Cancer risk in 680,000 people exposed to computed tomography scans in childhood or adolescence: data linkage study of 11 million Australians. BMJ
18. Berthel E, Pujo-Menjouet L, Le Reun E, Sonzogni L, Al-Choboq J, Chekroun A et al (2023) Toward an early diagnosis for Alzheimer's disease based on the perinuclear localization of the ATM protein. Cells
19. Al-Choboq J, Mathis T, Restier-Verlet J, Sonzogni L, El Nachef L, Granzotto A et al (2023) The radiobiological characterization of human and porcine lens cells suggests the importance of the ATM kinase in radiation-induced cataractogenesis. Cells
20. Aleci C (2020) From international ophthalmology to space ophthalmology: the threats to vision on the way to Moon and Mars colonization. Int Ophthalmol
21. Cucinotta FA, Manuel FK, Jones J, Iszard G, Murrey J, Djojonegro B et al (2001) Space radiation and cataracts in astronauts. Radiat Res
22. Blakely EA, Chang PY (2007) A review of ground-based heavy ion radiobiology relevant to space radiation risk assessment: cataracts and CNS effects. Adv Space Res
23. El Nachef L, Berthel E, Ferlazzo ML, Le Reun E, Al-Choboq J, Restier-Verlet J et al (2022) Cancer and radiosensitivity syndromes: is impaired nuclear ATM kinase activity the primum Movens? Cancers

24. Deschavanne PJ, Fertil B (1996) A review of human cell radiosensitivity in vitro. Int J Radiat Oncol Biol Phys
25. National Institutes of Health. National Cancer Institute. Common Terminology Criteria for Adverse Events (CTCAE) Version 4.03: June 14, 2010
26. Fertil B, Malaise EP (1981) Inherent cellular radiosensitivity as a basic concept for human tumor radiotherapy. Int J Radiat Oncol Biol Phys
27. Ferlazzo ML, Sonzogni L, Granzotto A, Bodgi L, Lartin O, Devic C et al (2014) Mutations of the Huntington's disease protein impact on the ATM-dependent signaling and repair pathways of the radiation-induced DNA double-strand breaks: corrective effect of statins and bisphosphonates. Mol Neurobiol
28. Combemale P, Sonzogni L, Devic C, Bencokova Z, Ferlazzo ML, Granzotto A et al (2022) Individual response to radiation of individuals with Neurofibromatosis type I: role of the ATM protein and influence of statins and bisphosphonates. Mol Neurobiol
29. Al-Choboq J, Ferlazzo ML, Sonzogni L, Granzotto A, El-Nachef L, Maalouf M et al (2022) Usher syndrome belongs to the genetic diseases associated with radiosensitivity: influence of the ATM protein kinase. Int J Mol Sci
30. Bodgi L, Granzotto A, Devic C, Vogin G, Lesne A, Bottollier-Depois JF et al (2013) A single formula to describe radiation-induced protein relocalization: towards a mathematical definition of individual radiosensitivity. J Theor Biol

Providing Efficient CBRN Protection Training for the Citizens of the Republic of Moldova

Vitalie Marinuta and Vitalie Stoian

1 Introduction

In the last period of the war of aggression of Russia against Ukraine, Kremlin officials and Russian opinion makers use more and more the threat using nuclear weapon. Most of the specialist agree that even a detonation of tactical nuclear weapon at surface-level in Ukraine, may not only cause significant casualties in the area of detonation, but also radioactively contaminate large territories, including the surrounding countries. As the Republic of Moldova is only few hundred kilometers from the war zone, the risk of being significantly affected in the event of the use of NBC weapons in Ukraine, especially in the southern part, increases substantially.

In this article the authors argue that even military threat, especially CBRN treat, is not significant at the moment for the Republic of Moldova, with increasing rhetoric of the use by Russian Federation of the nuclear weapons in the war of aggression against Ukraine, the Government has to invest more not only in conventional training of its regular and reserve military forces, but also to train troops and the population to be able to act accordingly in case of CBRN treat. One of the focal points in providing training to the civilian population as well as for the reservists should be experienced reserve and retired military officers and NCOs from the Military departments of the Universities as well as some of the NGOs that have experience in that domain.

V. Marinuta (✉)
Retired Officers Association, Chișinău, Republic of Moldova

V. Stoian
Military Training Department, Moldova Technical University, Chișinău, Republic of Moldova
e-mail: vitalie.stoian@mil.utm.md

F.-C. Cirstoiu et al., *Medical Response Strategy in Case of Radiation Emergency Caused by the War in Ukraine*, NATO Science for Peace and Security Series B: Physics and Biophysics, https://doi.org/10.1007/978-94-024-2266-5_12

2 Situation in Ukraine

On 24 February 2022 Russian Federation started the active faze of the invasion of Ukraine. At this moment Russia had already under its control about 7% of the Ukrainian territory as a result of the occupation in 2014 of Crimea and intervention in the Donbas region [1].

The invasion started with many missiles strikes on most Ukrainian cities. In few weeks Russian ground forces have already taken large areas in the east and south, had advanced to the suburbs of Kyiv and surrounded the port city of Mariupol. Ukrainian Armed Forces and civilians quickly organized and resisted to the invasion. Russian troops had poor logistic and they luck efficient leadership and motivation. International community had organized and provided Ukraine very fast with arms, artillery systems and ammunitions as well as financial support that helped not only to resist, but also quickly to change the situation.

Late summer of 2022 was a very difficult period for the Russian troops. Ukrainian forces were advancing in Kherson region in the south. The city had been Russia's biggest achievement in this invasion, and it was a danger to be lost. In addition, it was verry possible that many Russian units would be surrounded. In this situation, at least US administration thought that such a catastrophic loss could be a "potential trigger" for the use even of nuclear weapons by Russians.

In 2023, none of the parts have gained important success on the ground. The front line was almost 1000 kilometers and did not change much by the fall of that year. With significant help from the international community, in the summer of 2023, Ukrainians have prepared and mounted unsuccessfully a contra offensive. Nevertheless, Ukrainian successful sea operations forced Russian fleet to withdraw from its coast of the Black Sea. Many military specialists conclude that the war has entered into a faze of attrition, because none of the parts had a capability to win over other.

International Community, including NATO and EU countries, strongly condemn Russia's brutal war of aggression against Ukraine and its illegal annexation of the Donetsk, Luhansk, Zaporizhian and Herson regions of Ukraine. They also condemn the involvement of Belarus in Russia's military aggression. Since February 2022, the European Council and the Council of the European Union have met regularly to discuss the situation in Ukraine from different perspectives. EU leaders have repeatedly called on Russia to immediately cease military action, unconditionally withdraw all military forces and equipment from Ukraine and fully respect Ukraine's territorial integrity, sovereignty and independence. The EU has pledged to continue to show solidarity and provide support for refugees fleeing the war in Ukraine and for the countries hosting them. The EU is also coordinating with partners and allies, within the UN, OSCE, NATO and G7.

Ukraine is not a NATO member, but has been a "partner country" since it applied for an Action Plan for membership. All NATO countries are also in permanent dialog between then and with Ukraine and are helping to win the war by providing deferent type of arms, including air defense systems, tanks and airplanes. According to the declaration of NATO Secretary-General Jens Stoltenberg, NATO is not taking part in the conflict triggered by Russia but will support Ukraine as long as needed.

3 Nuclear Threat from Russian Federation

Since the very beginning of the active faze of the Russian invasion its rhetoric about using the nuclear weapons is getting stronger. In the first days of the invasion of Ukraine, Russian president Vladimir Putin transferred its countries nuclear forces into a state of high alert [2].

In September 2022 Russian president in one of its speeches mentioned that they will use all the means that are at their disposal to protect the country. This declaration was widely seen as a treat to use nuclear weapons, especially that he accused NATO of preparing its nuclear forces against Russia [3].

Over the time there were more declarations that could be classified as treats to use weapons of mass destructions, especially tactical nuclear weapons from former prime minister Dmitry Medvedev, foreign minister Sergey Lavrov, minister of defense Sergey Shoigu and other civilian and military high-ranking officials [4].

In the same period, Russian propagandists were circulating a false information about Ukraine using a dirty bomb. Some western analysts were concerned that these stories are for legitimization of Russia using tactical nuclear weapons. In October 2022, Sergei Shoigu, Russia's defense minister, contacted some of the defense ministers from biggest NATO countries (US, the UK, France and Turkey), telling them that the Kremlin was "concerned about possible provocations by Kyiv involving the use of a dirty bomb." Russia also sent an official letter to UN with the same allegations that Ukraine is building and planning to detonate a dirty bomb and blame that on Russia [5]. Administrations of many western democracies did not trust Russia on that issue. International reaction to the treats of the Russian Federation to use nuclear weapons was prompt. United States of America classified that action as "totally unacceptable". In May 2022, the US Senate held the "Hearing on Nuclear Readiness Amid Russia-Ukraine War". During the hearing Admiral Charles A. Richard stated that current Russian and Chinese nuclear stockpiles are larger than those of the US. In addition, he concluded that nuclear defense capabilities of US were operating at a minimal acceptable level of operational capacity [6].

U.S. President Joe Biden condemned Putin's threat to use nuclear weapons and urged allied U.N. leaders to reject Russia's invasion of Ukraine. Speaking at United Nations headquarters in New York City in September 2022, Biden accused the Kremlin of making "reckless" and "irresponsible" threats and said, "a nuclear war cannot be won and must never be fought." NATO Secretary-General Jens Stoltenberg, also mentioned that the alliance would remain calm and "not engage in that same kind of reckless and dangerous nuclear rhetoric as President Putin." Some high officials of the NATO countries also agreed that in case of use by Russia of nuclear weapons, the response should be devastating but not nuclear [7].

Minister of Foreign Affairs of Poland, Zbigniew Rau, has stated a NATO response should be "devastating", but non-nuclear. As respond to Russia basing the nuclear weapons in Belarus, which is perceived by Poland as an existential treat, its government asked US to deploy also nuclear weapons on its territory.

Some political and military analysts as well as writers even developed the list of possible targets of the nuclear attack in that invasion. Eric Schlosser, a writer for The Atlantic magazine suggests: "(1) a detonation over the Black Sea, causing no casualties but demonstrating a resolve to cross the nuclear threshold and signaling that worse may come, (2) a decapitation strike against the Ukrainian leadership, attempting to kill President Volodymyr Zelensky and his advisers in their underground bunkers, (3) a nuclear assault on a Ukrainian military target, perhaps an air base or a supply depot, that is not intended to harm civilians, and (4) the destruction of a Ukrainian city, causing mass civilian casualties and creating terror to precipitate a swift surrender—the same aims that motivated the nuclear attacks on Hiroshima and Nagasaki."

Officials from the countries neighboring the war theater should as a precaution take in consideration all the necessary measures to protect their population. In order do minimalize the casualties between civilian population, the governments need to inform and train it on how correctly to act in such cases. Having trained personal within civilian institution, at all levels of the administration and conducting periodically elementary trainings, aimed at creating certain behavior of the population, as well as informing timely and correctly the population about CBRN treats is crucial for saving people's lives.

4 CBRN Protection of Population in the Republic of Moldova

General Inspectorate for Exceptional Situations from the Ministry of Interior Affairs is main state authority responsible for providing information, training and civil protection, including the case of CBRN treats in the country. Civil protection of the Republic of Moldova is a system of measures and actions, undertaken within the state in times of peace and war, in order to ensure the protection of the population and property in conditions of natural and ecological disasters, epidemics or fires. The European Union and the Republic of Moldova at the end of September 2023 signed an agreement for Moldova to become a participating state in the EU Civil Protection Mechanism. According to the agreement, as a full member of the EU Civil Protection Mechanism, Moldova will not only be able to receive immediate support, but can also send assistance to countries struck by a disaster through the Mechanism, leading to a stronger and better coordinated crisis response in Europe and in the rest of the world [8].

In September 2022, Moldova's National Security Counsil (NSC) estimated that the risk of nuclear contamination for the Republic of Moldova is very low, but still exists. Those are external risk factors:

1. The use of nuclear weapons in the Russian Federation's war against Ukraine;
2. Incidents at nuclear power plants in neighboring countries, which are in the relative proximity to the Republic of Moldova (135–400 km): Ukraine—Rivne (275 km), Khmelnitsky (120 km), Chernobyl (350 km), Novo-Ukrainka (150 km), Zaporishe (450 km); Romania—Cernavoda (120 km); Bulgaria—Cozlodui (425 km).

As a result, NSC tasked the Government to reevaluate its contingency plans. The NSC recommended to the government also to check how prepared institutions and the population are to deal with the consequences of a possible nuclear disaster. At the same time, the council advised the authorities to launch an information campaign to inform citizens about protective measures in the event of a nuclear attack. Since then, unfortunately not much was done to inform the population of the Republic of Moldova about the treats of the CBRN, but there are permanent trainings, including at the international level, with the specialists of CBRN protection from the General Inspectorate for Exceptional Situations from the Ministry of Interior Affairs. The aim of the international field exercises is manly to create a dynamic support in joint interventions, where each team can contribute its own experience, methods and techniques in the process of liquidating the consequences of CBRN accidents. At the same time, the joint applications aim to improve the level of collaboration, effective communication and operational management between the teams involved.

In the Republic of Moldova, there are specially designed underground shelters where people could take cover in case of military conflicts, chemical or biological attacks, nuclear accidents or other exceptional situations. Fifty years ago, underground shelters were equipped with medical kits, masks, ventilation systems and power generation systems. When the USSR collapsed those facilities were abandoned. Today only few of them can be used as shelters in case of need.

One of the best maintained bunkers is that from the Moldova State University of Medicine and Pharmacy, which is located in the basement of a study block and is designed only for 25 people. This underground facility has a command post for the university administration and a protective area, enclosed by an armored door that closes tightly. The command post is equipped with a long table, chairs, disinfectant and masks, and several dozen bottles of drinking water. The area is equipped with internet and telephone access, protective suits, plastic and metal helmets, fire extinguishers, medical kits, masks and anything else is needed in case of danger. This buncher could serve as an example how should be organized other shelters in the Republic of Moldova [9].

Currently, Republic of Moldova does not have in its government institutions enough specialists that would have needed qualification to deal with CBRN treats. Besides few specialists from the General Inspectorate for Emergency Situations, that are directly responsible for dealing with CBRN treats there are some officers within chemical protection unit of the National Army and in Atomic Agency. Thus, one of the basic categories of specialists that have minimal necessary training in the Republic of Moldova to deal with the consequences of the use of CBRN agents and could be effectively used in those situations are trained reservists of the Armed Forces. Those officers and NCOs are active within military departments of the state Universities and in some specialized NGOs.

Citizens of Moldova are getting their specialized training in the framework of military conscription service, during military exercises with reservists, organized by the National Army and at the military departments of the Universities. In addition, in the Moldovan legislation there are provisions about organization of pre-military training on a voluntary basis, with medically fit recruits, with the aim of acquiring knowledge and training military and technical-application skills, developing patriotic attitudes and forming moral principles, as well as educating them as citizens. The organization of pre-military training centers and development of training programs are stipulated in regulations developed by the Ministry of Defense in cooperation with other central government authorities.

The military training of the reservists outside the Armed Forces is provided by the military departments of the public Universities. The training programs of military departments are approved by the Ministry of Defense. The technical, material and financial support of the military departments shall be provided by the Universities. Students attend the training course at the military department on a voluntary basis. After graduation from the full training course at the military department, depending on the needs of the Armed Forces and the results obtained in the military training, graduates of higher education institutions may be granted the rank of officer or sergeant in reserve. Citizens who have completed the full course of training at the military department but have not graduated from the educational institution may be granted the rank of sergeant in reserve.

In accordance with Moldova's legislation, military departments within Universities are organized based on the Constitution of the Republic of Moldova, the Law No.1245-XV of 18 July 2002 on the training of citizens for the defense of the homeland, the Law No 1244 of 18 July 2002 on the Armed Forces Reserve, Decision of the Government of the Republic of Moldova No 587 of 20.05.2003 on the approval of the Regulation on the activity of military departments in state higher education institutions, the Order of the Minister of Defense Education Ministers, the University Senate decisions and Rectors' orders.

One of the most efficient military departments is that of the technical University of Moldova. The Military Department was founded on the basis of the decision of the TUM Senate of 29.10.2002 and the Rector's Order no. 758-R of 18.11.2002. The main mission of the Department is the qualified training of sergeants and soldiers for the Armed Forces Reserve, with a theoretical training and practical skills appropriate to the respective specialty, able to carry out at any time military actions individually and at group / team level. The headquarters of the department is located in Chisinau municipality, Botanica sector. The military department also has subdivisions in the Centre and Rascani districts of the capital, as well as in the municipalities of Balti and Soroca. Since the creation of the department, some 26,500 people have graduated from the department's training courses, including 1098 women.

CBRN theoretical and practical training within the department is organized and conducted by former active officers (LTC of Colonels) with experience of more than 25–30 years in military units specialized in that domain. Education and training is organized using contemporary protection equipment and educational materials as well as new methods of teaching. For learning of this subject, the department allocates 12 academic hours, which is enough to develop the basic skills for efficient personal protection in case of need.

Another category of specialists that could be involved in the CBRN protection education of the population are members of military reserve officers' associations. One of this entities is the Association of Veterans and Reservists of the Republic of Moldova, which is a non-commercial and apolitical entity established in 2013 with the purpose of capitalizing on the experience and professional skills of military reserve and retired personnel, to contribute to the promotion of national and Euro-Atlantic security interests, values and objectives; of providing advice, education and training to central and local authorities, non-commercial organizations and other interested public entities in the Association's areas of expertise (strategic planning, security and defense, civil protection, unexploded ordnance, chemical and nuclear protection), as required. In 2014, the Ministry of Defense had signed a Cooperation Agreement with AVRRM, which set out the principles of cooperation in the areas of youth education, social protection of veterans and support in the training of reservists.

Over the years, the Association has initiated various projects to provide aid to veterans and their family members, conducted numerous educational-patriotic meetings in schools and high schools, supported citizens' actions against corruption, participated in campaigns to combat misinformation about Moldova's Euro-Atlantic path and is one of the civil society entities monitoring the process of restructuring the security and defense sector of the Republic of Moldova. Since the beginning of the war in Ukraine, the Association has been involved in various projects providing aid to refugees in the Republic of Moldova, collecting funds and basic necessities, donating blood for comrades in the Ukrainian Army, conducting training and education in the field of protection against unexploded ordnance (EORE). The members of the Association are former military and senior civilian employees from different components of the security and defense sector, scientists, civic activists and war veterans, who have a lot of experience in the field and are supporters of the fight against corruption and oligarchic structures and the Euro-Atlantic course of development of our society.

In summary, in current security situation in the region and with increasing threat of the use by Russian Federation of the nuclear weapons in the war of aggression against Ukraine, the government of the Republic of Moldova have to use the potential and the experience of the reserve and retired officers from the Military departments of the Universities and specialized NGOs to teach the population basic principles of protection against CBRN treats.

References

1. State Statistics Service of Ukraine (2024). https://ukrstat.gov.ua/
2. Putin signals escalation as he puts Russia's nuclear force on high alert, https://www.theguardian.com/world/2022/feb/27/vladimir-putin-puts-russia-nuclear-deterrence-forces-on-high-alert-ukraine. Last accessed in 11.12.2023
3. 'I'm Not Bluffing': Vladimir Putin Warns The West He Is Willing To Use Nuclear Weapons, https://www.huffingtonpost.co.uk/entry/vladimir-putin-says-he-is-willing-to-use-nuclear-weapons_uk_632aacf1e4b07198f013d40f. Last accessed in 11.12.2023
4. Medvedev raises specter of Russian nuclear strike on Ukraine. https://www.reuters.com/world/europe/russias-medvedev-warns-west-that-nuclear-threat-is-not-bluff-2022-09-27/. Last accessed in 11.12.2023
5. Exclusive: US prepared 'rigorously' for potential Russian nuclear strike in Ukraine in late 2022, officials say. https://edition.cnn.com/2024/03/09/politics/us-prepared-rigorously-potential-russian-nuclear-strike-ukraine/index.html#:~:text=In%20October%202022%2C%20Russia's%20defense,use%20of%20a%20dirty%20bomb.%E2%80%9D. Last accessed in 15.01.2023
6. Hearing on Nuclear Readiness Amid Russia-Ukraine War. https://www.c-span.org/video/?519954-1/hearing-nuclear-readiness-amid-russia-ukraine-war#. Last accessed in 12.12.2023
7. What if Russia Uses Nuclear Weapons in Ukraine? https://www.theatlantic.com/ideas/archive/2022/that 06/russia-ukraine-nuclear-weapon-us-response/661315/. Last accessed in 11.12.2023
8. 12 salvatori din Republica Moldova participă la Exercițiul internațional de teren CBRN în Polonia, https://www.dse.md/ro/communicate/12-salvatori-din-republica-moldova-particip%C4%83-la-exerci%C8%9Biul-interna%C8%9Bional-de-teren-cbrn
9. În căutarea adăpostului pierdut. Cum arată buncărele din Moldova și câte persoane pot adăposti. https://newsmaker.md/ro/in-cautarea-adapostului-pierdut/. Last accessed in 21.02.2024

Medical Units' Strategic Management in Case of CBRN Incidents in Republic of Moldova

Sergiu Cîrlan, Andrei Marfin, and Vasile Dumitraş

1 Introduction

CBRN emergencies can occur as a result of industrial disaster, occupational exposure, natural catastrophe, warfare, or acts of terrorism. Medical support is one significant component of nuclear and radiological emergencies.

The Zaporizhzhia Nuclear Power Plant (or Zaporizhzhia NPP) is the largest nuclear power plant in Europe and one of the largest in the world. Located in Ukraine, near the city of Enerhodar, this power plant has six nuclear reactors in operation. Currently, in the respective region, intense combat operations are underway.

In the event of a major accident or incident at the Zaporizhzhia Nuclear Power Plant, there would be significant risks to public health and the environment. These risks include:

1. Explosion and fire: A major accident could lead to an explosion or fire inside the reactor, releasing radioactive materials into the environment. This could lead to contamination of the air, soil, and water in the region.
2. Spread of radiation: Radioactive materials released into the atmosphere could be carried by the wind and could affect extensive areas around the nuclear power plant. This could endanger the health of the population in the affected areas.

S. Cîrlan (✉)
"Alexandru cel Bun" Military Academy, Chişinău, Republic of Moldova
e-mail: sergiu.cirlan@usmf.md

A. Marfin
Outpatient Center of Ministry of Defence, Chişinău, Republic of Moldova

V. Dumitraş
Department of Military and Disasters Medicine, "Nicolae Testemitanu" State University of Medicine and Pharmacy, Chişinău, Republic of Moldova

F.-C. Cirstoiu et al., *Medical Response Strategy in Case of Radiation Emergency Caused by the War in Ukraine*, NATO Science for Peace and Security Series B: Physics and Biophysics, https://doi.org/10.1007/978-94-024-2266-5_13

3. Effects on human health: Exposure to radiation can cause serious illnesses such as cancer and other health conditions. Affected areas may require evacuation and special measures to protect against radiation.
4. Impact on the environment: Contamination with radiation could affect local ecosystems and could have long-term consequences for biodiversity and the quality of the surrounding environment.

In accordance with the military strategy approved in October 2018, the National Army of the Republic of Moldova has the following missions:

1. Ensuring national defense by preventing, countering, and repelling military threats and aggressions directed against the Republic of Moldova;
2. Ensuring participation in international peacekeeping missions/operations;
3. Supporting civilian authorities in Emergency situations.

In the event of a possible accident at this nuclear power plant, supporting civilian authorities will become the primary mission of the National Army.

Results According to IAEA Safety Standards Series No. GSR Part 7, Preparedness and Response for a Nuclear or Radiological Emergency, an emergency is: "A non-routine situation or event that necessitates prompt action, primarily to mitigate a hazard or adverse consequences for human life, health, property or the environment. This includes nuclear and radiological emergencies and conventional emergencies such as fires, releases of hazardous chemicals, storms or earthquakes. This includes situations for which prompt action is warranted to mitigate the effects of a perceived hazard."

Nuclear and radiological events—an emergency in which there is, or is perceived to be, a hazard due to: energy resulting from a nuclear chain reaction or from the decay of the products of a chain reaction; radiation exposure. Rephrasing that, we need to distinguish between Irradiation vs. Contamination.

Irradiation person is irradiated when they are "exposed" to ionizing radiation in much the same way a person is "exposed" to light when someone shines a flashlight on them. In the case of irradiation, there is no material transferred. This means that an irradiated patient has no radioactive material on them and poses no radiological hazard to the treatment team.

Contamination when people have radioactive materials on/in them, they are said to be contaminated.

Any operation acting in a region with nuclear reactors should consider the possibility of a contingency plan in the event of a nuclear incident or accident. The impact of such an event: potential traumatic injuries either during the initial incident or in response; high dose or dose rate exposures due to the nuclear process; exposure to a range of radioisotopes (fission products), many of which will be highly potent during the first few days following an event. Most will be beta emitters with associated gamma emissions; exposure to radioiodine (~2% of fission products) and potential for stable iodine to be used as a radiological MedCM; risk of internal contamination either by inhalation or ingestion either directly or through the water and food chain; risk of fallout from airborne nuclear material; risk of direct irradiation (shine) from settled nuclear material.

A review of past accidents involving radioactive sources identified the following three generic exposure scenarios involving radiation injury to an individual as being typical of many situations. These generic scenarios may be used to estimate potential radiation exposures which may occur if control of sources is lost:

1. External exposure to an individual from a source in very close proximity. A radioactive source can be lost or stolen. One scenario that has occurred repeatedly is when a person places an unshielded radioactive source in their pocket. There are various reasons why a person may do this, including theft, curiosity about an unknown object, or for the convenience of carrying it to another location. The individual involved may be an employee of the facility or a regular member of the public, which can lead to significant doses for one or more individuals.
2. External exposure from an unshielded source. It is possible that once control of a source has been lost, it may irradiate workers or members of the public without the knowledge of those involved. Again, gamma radiography provides an example: after making an exposure, the source may become jammed and the operator(s) and other persons may be unintentionally exposed, resulting in significant doses to several persons.
3. Exposure following rupture of source casing. If a source which is not controlled becomes ruptured, the radioactive material may be dispersed, resulting in contamination of equipment or individuals; exposures from inhalation of radioactive material, inadvertent ingestion of radioactive material, and contamination of the skin; and external exposure from the spillage.

The goals of the medical response to a nuclear or radiological emergency are as follows: to save lives and perform required emergency medical procedures; to treat radiation injuries and injuries resulting from an emergency situation; to perform required public health actions, including providing public advice and counseling, and long term medical follow-up of exposed persons.

Requirement 12 of GSR Part 7 states: "The government shall ensure that arrangements are in place for the provision of appropriate medical screening and triage, medical treatment and longer term medical actions for those people who could be affected in a nuclear or radiological emergency.

Based on the analysis of previous emergencies, the medical aspects of the preparedness and response to radiation emergencies have demonstrated gaps that need to be addressed. Specific weaknesses include the following: lack of training of medical professionals in recognizing radiation induced injuries and understanding the specifics of treating these patients; lack of appropriate medical assessment relating the severity of radiation dose with the clinical manifestations; lack of knowledge of the processes for notifying the national authorities; lack of availability of a sufficient number of medical staff to deal with the projected number of victims affected in a radiation emergency; lack of provisions in the national emergency plan for promptly requesting emergency assistance from international organizations (under the Convention on Assistance in the Case of a Nuclear Accident or Radiological Emergency) for dealing with victims; lack of established criteria for determining the groups which have been highly exposed and should be subject to long term medical follow-up to detect the early appearance of cancer and other effects.

Emergency medical response needs to be planned and organized in accordance with the potential consequences of different radiation emergencies. The basic medical preparedness requirements for responding to a radiation emergency will consider: the evaluation of possible types of emergency according to the local circumstances; training of health care professionals; the provision of simplified dosimetry services (a dose rate meter at the 'pre-hospital' stage); capabilities for detecting radiation and containing the spread of contamination; basic preparations for receiving exposed or contaminated patients; simple medical procedures to be undertaken by non-specialized staff.

In addition, the planning of the medical response needs to include: medical and radiological triage; an initial medical examination; clinical, laboratory and dose assessments; designation of medical institutions for management of patients; treatment in specialized medical centres (for severe radiation injuries); medical follow-up.

In planning the medical response to nuclear emergencies, decontaminating patients before admission to medical treatment facilities is of paramount importance. Contamination with radionuclides may be external or internal, or both. Contaminated patients can have radioactive material deposited on skin surfaces, in wounds or internally.

External contamination occurs when radioactive material from an unsealed or damaged source is deposited on the surface of the skin or clothing. It is detected through direct monitoring of the skin and clothing. Patients who exhibit external contamination of the skin, clothing, or excreta can contribute to the spread of contamination, thus special precautions are necessary to prevent this.

For those in need of urgent medical care and immediate transport to hospitals or other medical centers, priority will be determined based on their medical condition, without taking contamination into consideration. When medical personnel adhere to standard universal biosafety measures in treating patients, this will contribute to contamination control.

Individual monitoring will not disrupt medical interventions aimed at stabilizing a person's condition or transporting patients with life-threatening injuries. Radiological assessments at the scene of the incident can be conducted in collaboration with medical personnel only for hemodynamically stable patients, depending on the conditions encountered at the scene.

If patients are externally contaminated, their clothing should be promptly removed, if it has not already been removed before arrival at the hospital, unless medically contraindicated. Care must be taken to prevent the spread of contaminants embedded in or on the clothing. Place the clothing and any accompanying bedding or blankets in sealed plastic bags, label them, and store them properly for radiological analysis.

Bibliography

1. Preparedness and response for a nuclear or radiological emergency, IAEA safety standards series no. GSR Part 7, IAEA, Vienna, 2015
2. AMedP-7 (2018) Medical management of CBRN casualties, Edition A, version 1. NATO, Brussels
3. International Atomic Energy Agency, World Health Organization (1998) Diagnostic and treatment of radiation injuries, Safety reports series nr.2. IAEA, Vienna
4. International Atomic Energy Agency, World Health Organization (2000) Categorization of radiation sources, IAEA-TECDOC-1191. IAEA, Vienna
5. International Atomic Energy Agency (2020) Medical management of radiation injuries., Safety report series No. 101, Vienna
6. International Atomic Energy Agency (1998) The radiological accident in Tammiku. IAEA, Vienna
7. International Atomic Energy Agency (2000) The radiological accident in Istanbul. IAEA, Vienna
8. International Atomic Energy Agency (2005) Generic procedures for medical response during a nuclear or radiological emergency. EPR-Medical, IAEA, Vienna

Scientific and Evidence-Based Support in Emergency Preparedness and Resilience Strengthening, Under Volatility and Uncertainty

Vamanu Bogdan-Ioan, Petre Min, Acasandrei Valentin-Teodor, and Pavelescu Alexandru Octavian

Abstract In the context of the crisis generated by the Russian-Ukrainian conflict, the situation of the five (four operational nuclear power plants) on Ukrainian territory has become a point of major public interest and a prominent sensitive issue for national decision-makers and response personnel. Since the beginning of the conflict, two research-oriented teams from the Horia Hulubei National Institute for R&D in Physics and Nuclear Engineering (IFIN-HH) in Bucharest, Romania and the National Commission for Nuclear Activities Control (CNCAN) have carried out a series of comprehensive situation radiological forecasting activities. The aim was to provide a plausible, well-founded, objective and scientifically defendable assessment of Romania's territorial vulnerability to a hypothetical radiological accident at any of the nuclear power plants in Ukraine. Evaluations were performed using two distinct platforms of dose projection and radiological assessment to ensure consistency of results. The scenarios were built on the assumption of loss of coolant event followed by a release of radioactive material in the atmosphere resulting from damage to the reactor core and spent fuel pool. The paper provides an overview of the work, addressing key topics such as situational awareness, expert judgment, underlying assumptions, scenario development, and methods for effectively producing and communicating meaningful results. It also highlights some particularly significant and interesting findings and concludes with a few thoughts and observations.

V. Bogdan-Ioan (✉) · A. Valentin-Teodor · P. Alexandru Octavian
National Institute for R&D in Physics and Nuclear Engineering, IFIN-HH, Magurele Ilfov, Romania
e-mail: bvamanu@nipne.ro

P. Min
National Commission for Nuclear Activities Control, Bucharest, Romania

F.-C. Cirstoiu et al., *Medical Response Strategy in Case of Radiation Emergency Caused by the War in Ukraine*, NATO Science for Peace and Security Series B: Physics and Biophysics, https://doi.org/10.1007/978-94-024-2266-5_14

1 Introduction

Conflicts always involve volatile and uncertain situations. Whether called *special military operation* or full-blown war, any military action can result in unforeseen and undesirable consequences. Furthermore, nowadays more than ever, due to the influence of social-media and other communication channels, strategic objectives of the parties are pursued not only through traditional means but also through 'soft' tactics like disinformation, half-truths, panic-induction, and fake news.

In both war and peace, nevertheless, the potential impact of nuclear accidents has been a topic of debate and a significant concern for the public. Starting February 24, 2022, it was somehow predictable that the situation of the five (four operational) nuclear power plants (NPPs) on Ukrainian territory would become a major point of interest for the public and a prominent sensitive issue for national policymakers and emergency responders. And the reality unfolding just confirmed these assumptions.

The Russian-Ukrainian conflict also posed a new challenge for Romania: a war at borders that could associate, in addition to the economic, social, and humanitarian crises, a possible nuclear crisis with consequences on the national territory.

It is in this context that, from February 24, 2022 onwards, teams from the 'Horia Hulubei' National Institute of Research and Development for Physics and Nuclear Engineering (IFIN-HH) and the National Commission for Nuclear Activities Control (CNCAN) conducted a series of comprehensive radiological situation forecasting activities, with the main objective to be ready to answer in a plausible, well-founded, objective, un-biased, and scientifically defendable manner to any concerns expressed by policy-makers, the national emergency system, and the public alike. Both teams are part of the CNCAN Working Group for the Risk Assessment in case of Nuclear and Radiological Emergencies (GLERURN), an entity established in 2016 and accommodating experts from various fields and having as main objective to provide scientific support in the emergency decision making process. The assessments were oriented towards identifying and monitoring the vulnerability of the national territory to a hypothetical radiological accident at any of the Ukrainian NPPs. More specifically, we aimed at providing plausible answers to the following topics: what may happen, where it may happen, would it affect us (Romania), and if so, where and to what extent.

The assessments were carried out using two distinct dose-projection and radiological assessment platforms: jRodos (Real-Time Online Decision Support System for the Management of Nuclear Emergencies) and CBRNE-Software, an IFIN-HH in-house solution, to ensure consistency of the results. In the absence of pertinent and confirmed information, several scenarios deemed plausible were developed based on situational awareness and expert judgment. Appropriate accident source terms were constructed using methods validated and approved by international organizations in the field, or referred to, by well-known design-based accident safety studies.

The endeavor resulted in more than 75 reports in total, including 50 daily situational reports covering all Ukrainian NPPs from February to May 2022, and over 25 reports specifically focused on individual NPPs between August and December 2022. Some of the reports were also used to substantiate and support public communication through mainstream media channels, both by CNCAN and IFIN-HH representatives.

In the sequel, the paper offers an overview of the undertaking, addressing key topics such as situational awareness, expert judgment, basic assumptions, scenario development, and the methods of producing and effectively communicating meaningful results.

2 The Scenarios

Ukraine is one of the most nuclear-dependent countries in the world, with nuclear energy accounting for more than 50% of electricity production. By the time the conflict started, four nuclear power plants (Khmelnytskyi, Rivne, Zaporizhzhia and South Ukraine) were in operation, summing up to 15 (13 VVER-1000, 2 VVER-440) reactors located in Volhynian and South Ukraine regions (Fig. 1). Of utmost importance is Zaporizhzhia NPP, the largest nuclear power plant in Europe, with a total installed capacity of over 13 GWe and supplying more than 20% of Ukraine's energy in 2019 [1]. In addition to the operational sites, Chernobyl-site hosted five units permanently shut down and the New Safe Confinement shelter of the affected Unit 4.

Since day 1 of the conflict the nuclear sites have been involved in military operations. On February 24, Ukraine informed the International Atomic Energy Agency (IAEA, Vienna) that Russian forces had taken control of all facilities of the Chernobyl nuclear power plant. According to various sources [2, 3], the moving of heavy military vehicles has raised radioactive dust and resulted in a spike of gamma radiation levels in the Chernobyl region 28 times above normal. Ukrainian authorities also claimed the Russian army is responsible for fire in the Chernobyl

Chernobyl	420 km
Khmelnitskiy	230 km
Rivne	345 km
South-Ukraine	260 km
Zaporizhzhia	490 km

Land	288.6 km
Rivers	292.2 km
Black Sea	33 km

Fig. 1 Geographical position of Romania in relation to Ukraine's NPPs and the borders area

exclusion zone, up to 37 hectares of forest having been swept by fires [4]. Control of the site was returned to Ukrainian personnel on 31 March.

On 4 March 2022, Russian forces occupied Zaporizhzhia NPP, the fighting resulting in a projectile hit of a training building; the reactors were not affected, and no environmental release of radioactive material occurred [5]. If these events were directly involving nuclear installations, other events during the unfolding of the military actions proven to be detrimental to the safety of operations of all of the NPPs on the Ukrainian territory. The attacks on the power infrastructure, for instance, lead to widespread blackouts resulting in lack of external power supply to all four of the country's nuclear plants [5]. Therefore, whether directly in the action zone or at hundreds of kilometers from the front-line, safety and security of any NPP is jeopardized in this uncertain, complex and intricate reality (Fig. 1).

Compared to its neighbors, Romania has the longest border with Ukraine with a total length of 613.8 km [6]. It consists of land, river, and sea border, of which 288.6 land km, 292.2 km by rivers and 33 km by the Black Sea and it is part of the external border of NATO and the European Union. In respect to the distance between the Ukrainian NPPs and national territory, this varies from 230 km (Khmelnytskyi) to 490 km (Zaporizhzhia). Considering the 1986 Chernobyl accident extent and scale, one may easily notice that Romania is not out of range of impact if any of the aforementioned NPPs are affected. This proximity and the—still live in public and policy-makers memory—past experience, naturally enhanced the anxiety and concerns in society as a whole, from the first days of the conflict.

Furthermore, the precarious situation of the NPPs in the context of the ongoing conflict provides fertile ground for misinformation and fake news. Narratives of the kind could rapidly spread across social media and other platforms, often with sensationalist headlines and unverified claims, which can exacerbate fears and induce widespread panic among the public. Given the inherent and, as mentioned, natural fear associated with nuclear incidents, false reports of radiation leaks, plant meltdowns, or sabotage could quickly gain traction, outpacing official communications, creating an environment of uncertainty and fear easily exploited by malicious actors aiming to destabilize public morale or manipulate political situations.

It became, therefore, crucial in this environment for authorities and the media to provide timely, accurate information to counteract the spread of disinformation and create a sufficient level of awareness and preparedness given the uncertainties and volatile situation at hand. Consistent to our mission, we strived to bring in a substantive to this endeavor to the best of our knowledge and expertise.

From a task-oriented perspective, the activities performed were focused on the following key aspects:

- Derive and draw plausible event scenarios (what can happen).
- Build the source terms (what can get out).
- Set-up simulations (other assumptions on release pathway, duration, etc.)
- Run the assessments (what and how much can be affected)
- Validation of the results (minimize errors in the assessment process)
- Prepare a situation report and communicate to the authorities.

3 The Scenarios Narrative

Event scenarios have been outlined by an analysis of the situation in Ukraine at that time and historical considerations. The outcome highlighted several essential aspects that had to be considered when developing the most probable event scenarios.

First, it had to be considered whether military operations were taking place on site or in the surroundings, and whether these actions resulted in the intentional or accidental hitting of the nuclear facility. Chernobyl fighting took place in day 1, yet the facility itself was not directly involved. Later in 2022, unknown at the time of this analysis, the military operations near Zaporizhzhia NPP, did however result in a projectile hitting a training building on the site near Unit no. 2. The reactors were not damaged, and no radioactive material was released into the environment. However, the IAEA reported many safety violations at the Zaporizhzhia NPP, as the physical integrity of the plant was not respected in the fighting [7]. Therefore, based on the known facts, one may assume with sufficient confidence that even if military operations took place on site and in the surroundings of the NPPs, the facilities themselves would not be targeted, and any incident would be unintentional.

Another important aspect is looking for precedence that nuclear facilities and especially nuclear power plants have been targeted before. The answer is yes. Nuclear installations have been targets during military air strikes, occupations, invasions, and campaigns. In the last four decades, there have been several attacks on nuclear facilities [8], yet none of these attacks resulted in significant radiation release. And the reason behind is that whether talking about Iran-Iraq war in the '80s, the First Gulf War in the '90s or other, more recent military operations on nuclear facilities, never these actions have been against *operational* installations. When talking strictly to NPPs, all these attacks were against reactors under construction or 'close to start-up'.

Therefore, the key assumption for the task at hand was to consider as event initiator an *unintentional* action resulted by mistake during an operation.

The next step in the logical framework was to identify, based on the initial assumption, *the pathways of radioactive release into the atmosphere*. Based on expert judgement, the most vulnerable to radioactive release into the atmosphere have been considered *the reactor core* and *the spent fuel pool*. While a direct hit of the reactor envelope is unlikely to produce actual damage to the core itself, the spent fuel pools are more vulnerable to such an event, since less or no structural protection exists in this case. A potential issue that may pose a threat to the reactor core, though, is the loss of cooling capability due to e.g. the incapacitation of the auxiliary installations. The most plausible scenario in which a catastrophe could occur is that—if the nuclear reactors and/or spent fuel pool are damaged the coolant is effectively lost, that leading to a Loss Of Coolant Accident (LOCA) type event [9]. This is also the assumption made in our scenario, namely supposing that an unexpected shell strikes and damage the reactor and temporarily disables all backup cooling safety systems. This would entail an emergency featuring a reduced level of water leaving the fuel rods uncovered for a certain period, which would result in a significant radioactive release, off one or more defense in depth structures.

The observations before were the basis of two scenarios considered for the following assessments. The narrative of the considered scenarios involved the *unintentional hit* of the active zones of the nuclear installations, followed by the *release of radioactive material into the atmosphere*. Two such release pathways were considered: the reactor itself—*reactor scenario* and spent fuel pool—*spent fuel pool scenario*.

In addition, a third scenario has been considered for Chernobyl, with an unintentional hit damaging the NSC of Unit 4 resulting in the atmospheric release of a fraction of the existing inventory.

4 Simulations Setup

4.1 Source Terms Estimation

The source terms have been inferred based on the assumptions of the considered scenarios, in the absence of a source term confirmed by the operator of the nuclear objective. The determination of the source terms derives from the methodology accredited by the U.S. nuclear regulatory body—U.S. Nuclear Regulatory Commission adopted in nuclear accident response manuals [10, 11, 12], assimilated and recommended by the technical document series IAEA TECDOC of the International Atomic Energy Agency [13]. Known sometimes as 'The Four-Factors Rule' (4FRule), this is a simplified solution, practical and providing sufficient degrees of confidence in making decisions to respond to abnormal nuclear events [14]. For each of these scenarios, using the dedicated module of the CBRNE-Software platform, medium severity source terms were built considering a VVER-1000 reactor. An additional source term has been inferred for Chernobyl NSC, using the available information [15] about the inventory in the confinement. Table 1 contains the total activity and number of radionuclides considered, for each scenario.

Table 1 Source terms type for each scenario

Scenario	Accident type	Total activity (Bq)	Number of radionuclides
Reactor	Containment bypass, gap release	4.63E+17	17
Spent pool	Spent fuel pool gap release, 1 batch affected, no fire	5.79E+17	17
Chernobyl NPP	Un-categorized	2.22E+27	25

Table 2 The scenarios assumptions considered

Scenario	Type of accident	Release duration (h)	Release height (m)	Release time (local time)	Monitoring time
Reactor	Containment bypass, gap release	1	10	11:00 AM	24 h
Spent fuel pool	Gap release, 1 batch affected, no fire	4	2	11:00 AM	24 h
Chernobyl	Un-categorized	4	10	11:00 AM	24 h

4.2 The Release Assumptions

The release assumptions considered for the scenarios are given in Table 2. The assessments were performed daily starting at 11:00 a.m. over a 24-h monitoring interval, using the specific meteorological conditions for that day.

4.3 The Meteorological Data

Two meteorological data sources have been utilized during the assessments, specific to each of the radiological impact assessment tools employed, JRodos and CBRNE Software. Each software platform has different ways of collecting raw weather data.

In JRodos, weather models can be national weather data online from weather towers or numeric data for global weather forecast such as those from the United States National Center for Environmental Predictions (NOAA-NCEP) available through the NOMADS online service [16]. For the specific object of this exercise, the latter have been used.

CBRNE Software uses primary meteorological forecast data on intervals between 8 and 72 h with 1 h time-resolution, acquired in real time from public weather services online. The weather data thus obtained is further amended with the computation of the atmospheric stability classes (Pasquill A-F). As providers of weather data, CBRNE Software uses three public sources for redundancy: World Weather Online (WWO) [17], VentuSky [18] and FreeMeteo [19]. WWO has been used as primary data source throughout this exercise.

5 The Assessment Tools

As previously mentioned, the radiological impact assessments have been performed using two dose-projection decision support systems: JRodos [20] and CBRNE Software [21].

JRodos (Realtime Online Decision Support System for nuclear emergency management) is a mature, comprehensive software package developed and maintained by a consortium of scientific research institutions in Europe and deployed on several national emergency centres around the World. The platform aims at providing a procedurally and data-consistent support for off-site emergency management at local, national and regional levels at times following a nuclear accident that cover the early (4–7 days), as well as the intermediate and long (ingestion) phases (months, years) in the development of the event, the economic consequences considered.

CBRNE Software system is an in-house solution developed and maintained at IFIN-HH. The platform has been developed as part of a research project in the field of national security and safety [21] and represents a MS&V (Modelling, Simulation and Visualization) application designed for anticipatory, prognostic assessment of CBRNE events (chemical, biological, radiological, nuclear, and explosive), comprising functions for diagnosis of effects and consecutive recommendation of response measures for an open and scalable variety of event scenarios.

6 The Assessments

The radiological situation for all four active nuclear power plants in Ukraine (Khmelnitsky, Rivne, South Ukraine, and Zaporizhzhia) as well as the decommissioned Chernobyl Power Plant, has been assessed for several months from the start of the conflict.

The assessments were performed daily starting at 11:00 a.m. over a 24-h monitoring interval, using the specific meteorological conditions for that day.

Both decision support software systems have been employed, the independent results being inter-compared for consistency, thus ensuring sufficient confidence in the prognostic results. The comparison involved evaluating the similarity in plume patters, coverage, and extent, as well as dose values both near the source and at selected locations. If significant discrepancies between the assessment results were found, the process was restarted. This would follow a thorough review of the simulation scenario setup (if the discrepancies were due solely to numerical results) or of the weather data acquisition and pre-processing (if the issue lay elsewhere). The outputs were considered '*consistent*' if the numerical results were within two orders of magnitude from each other.

A pre-defined report template was created to ensure assessment results were communicated in a unified and consistent manner. The report structure consisted of the following sections: Scenario Assumptions, Scenario Description, Executive Summary, and Potentially Affected Areas, supported by tables and situation maps. Detailed assessment data was included in an annex, which covered source terms, situation forecasts, and a comparison of results.

It is important to emphasize that the visual and numerical results presented in the reports represented protective quantities (i.e. equivalent and effective doses) aimed at preemptive or immediate intervention. These included Total Effective Dose Equivalent (TEDE), Total Acute Lung Dose (TALD), Total Acute Bone Dose (TABD) and Total Thyroid Dose (TTHD)—all measured in millisieverts (mSv).

As a result, the corresponding situational maps displayed the affected areas based on the projected TEDE (mSv), calculated from the time-integrated concentration over the monitoring period, which was considered the most critical dose for protective actions (Fig. 2).

Another important aspect to note is that, based on our analyses, the effects of a radiological accident in Ukraine would not be limited to Romania or a specific neighboring country. Naturally, the greatest impact would be at the local scale, near the event. However, other countries would also be affected to varying degrees, with the determining factor being the meteorological conditions at the time of the incident.

The assessments resulted in 95 reports in total, including daily situational reports covering all Ukrainian NPPs from February to May 2022, and other reports specifically focused on individual NPPs between August and December 2022. While the core of the assessments (February to May) were conducted with the scenarios, assumptions and workflow presented here, the other have been tailored (scope and source terms) to reflect the situation and concerns at date. Therefore, some of the assessments were addressing the Zaporizhzhia, other South Ukraine. The reports were also used to substantiate and support public communication through mainstream media channels, both by CNCAN and IFIN-HH representatives.

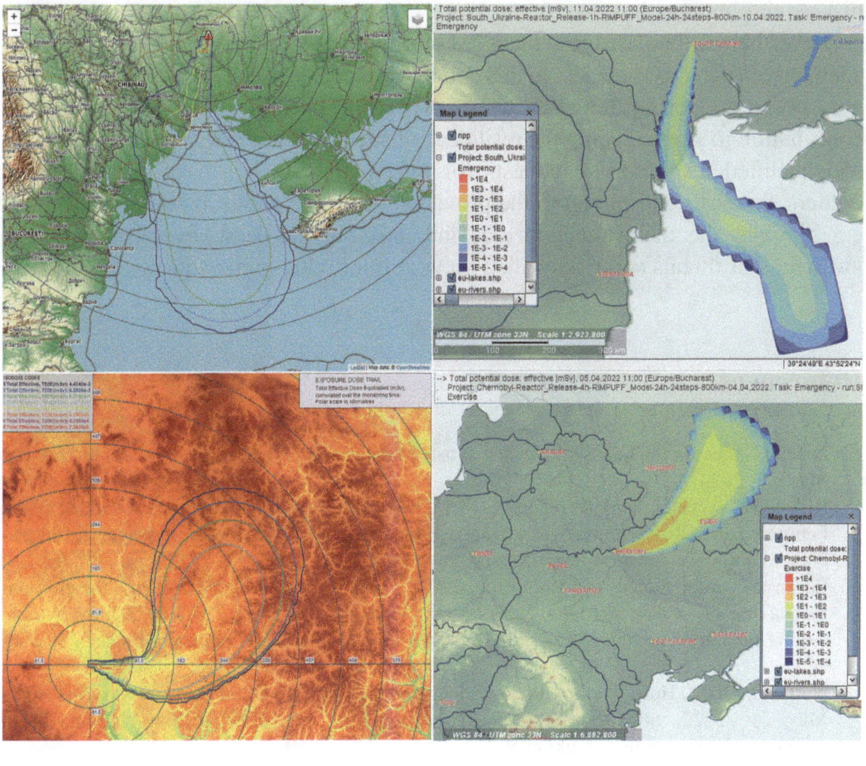

Fig. 2 Situation maps resulting from the evaluations of both software (left) CBRNE situation maps; (right) jRodos situation maps

7 Remarks and Conclusion

The military conflict in Ukraine has brought the issue of nuclear safety and security back into the spotlight. A response to legitimate concerns about *where and how it is happening* and *whether or not we will be affected,* has been demanded by the governing authorities, the media, and the population. This challenge was inevitably faced by the IFIN-HH and CNCAN teams involved in the multidisciplinary effort to identify methods for assessing, preparing for and response to such an event or to other nuclear and radiological emergencies of any kind.

Using two different decision support systems and building plausible event scenarios, a comprehensive analysis was carried out with the main purpose to identify the potential areas of influence and radiological consequences on the territory of Romania generated by the Ukrainian nuclear power plants, as a result of military actions.

Based on this analysis, the following results were obtained: each of the nuclear power plants located in the conflict zone may be vulnerable to a radiological accident. This may lead to the possibility of the national territory being affected, most probably in the eastern part of Romania. According to the analysis of different possible event scenarios, the radiological consequences in the potentially affected areas, in terms of doses to the population and the environment, are below the limits that require intervention.

Nevertheless, preparedness plays a key role in the effective and efficient management of such emergencies. In this context, we consider it relevant to play our part not only in supporting informed decision-making, but also in making our modest contribution to strengthening the resilience of society as a whole, by providing rapid and qualified responses to requests for information on behalf of societal stakeholders, thereby contributing to the provision of accurate information to the public, e.g. through the mainstream media, and, last but not least, by raising societal awareness of threats and risks with a view to improving preparedness.

References

1. Federal Office for the Safety of Nuclear Waste Management. Nuclear power plants in Ukraine. March 1, 2024
2. REUTERS Agency. Ukraine reports higher Chernobyl radiation after Russians capture plant. February 25, 2022. https://www.reuters.com/world/europe/ukraine-nuclear-agency-reports-higher-chernobyl-radiation-levels-due-heavy-2022-02-25/
3. European Union. Voices of young European journalists. Ukraine and the others: the environmental impacts of war. 30 March 2022. Last updated on Wednesday, 22/02/2023. https://youth.europa.eu/year-of-youth/young-journalists/ukraine-and-others-environmental-impacts-of-war_en
4. Nuttall, P (2022) How Vladimir Putin weaponized the environment in Ukraine. New Statesman May 21, 2022. https://www.newstatesman.com/long-reads/2022/05/how-vladimir-putin-weaponised-environment-in-ukraine

5. World Nuclear Association. Ukraine: Russia-Ukraine War and Nuclear Energy, Updated October 16, 2023. https://www.base.bund.de/EN/ns/country-information/ukraine_node.html
6. Wikipedia. Romania–Ukraine border. March 1, 2024. https://en.wikipedia.org/wiki/Romania%E2%80%93Ukraine_border
7. Nuclear Safety and Security in Ukraine, Update 93—IAEA Director General Statement on Situation in Ukraine, 23 August 2022
8. Carlson, J (2022) Prohibition of military attacks on nuclear facilities. Vienna Center for Disarmament and Non-Proliferation (VCDNP). September 12, 2022. https://vcdnp.org/prohibition-military-attacks-on-nuclear-facilities/
9. World Nuclear Association. Safety of nuclear power reactors. March 2022. https://world-nuclear.org/information-library/safety-and-security/safety-of-plants/safety-of-nuclear-power-reactors.aspx
10. McKenna TJ, Trefethen JA, Li Z (1995) IRTM—international response technical manual, vol 1, 2 and 3, Rev.1. U.S. Nuclear Regulatory Commission
11. McKenna TJ, Jackson K, Trefethen JA, Bores R, Struckmeyer R, Kuzo G, Hogan R, Jolicoeur J, Quissell J, Gant K, Burnett B, Aguilar-Torres O, Athey G, Grove L, Hopkins R (1993) RTM93—response technical manual, vol 1, Rev. 3. US Nuclear Regulatory Commission, NUREG/BR-0150
12. McKenna TJ, Trefethen JA, Gant K, Jolicoeur J, Kuzo G, Athey G (1996) RTM96—response technical manual, vol 1, Rev. 4. U.S. Nuclear Regulatory Commission, NUREG/BR-0150
13. International Atomic Energy Agency, IAEA, Vienna (1997) Generic assessment procedures for determining protective actions during a reactor accident. IAEA-TECDOC-955, ISSN 1011-4289
14. Vamanu DV, McKenna TJ (1996) ROBOT—rule-oriented basic operational tool. U.S. Nuclear Regulatory Commission Users' Guide, Rev 2
15. OECD. CHERNOBYL—Assessment of Radiological and Health Impacts (2002). https://www.oecd-nea.org/upload/docs/application/pdf/2022-01/3508-chernobyl_2022-01-05_11-11-9_843.pdf
16. National Oceanic and Atmospheric Administration (NOAA) Operational Model Archive and Distribution System (NOMADS)
17. World Weather Online Service. https://www.worldweatheronline.com/
18. Ventusky Web Application. https://www.ventusky.com/
19. FreeMeteo Service. https://freemeteo.com/
20. RODOS: Realtime Online Decision Support System for nuclear emergency management, Last modified: March 25, 2015
21. CBRNE Software Platform. Developed in the framework of "Integrated system for rapid intervention in CBRNE events" Project. SOL2017-7 2017-2020

Advocating for a Resilient Future: Key Takeaways for: "Regional Strategy for Medical Response as Part of the Disaster Management in Case of Radiation"

Niculae Iancu

The significance of "Regional Strategy for Medical Response as Part of the Disaster Management in Case of Radiation" this workshop was deeply emphasised by President Volodymyr Zelensky's speech at the UN General Assembly in New York, which fortuitously took place on the Bucharest event's inaugural day. Zelensky vividly highlighted how Russia, in its aggression against Ukraine, is harnessing essential resources, notably nuclear energy, food, and fossil fuels. The worrying perspective that Russia might convert Ukrainian nuclear sites into dirty bombs has amplified existing fears. The core message emerges openly from the labyrinth of concerns: The fear of a nuclear showdown, accentuated by Russia's calculated strategies, has repercussions far beyond Ukraine's borders. The actions of a dominant state like Russia can, and likely do, influence how the world responds to forceful actions and land grabs by other assertive nations. Therefore, President Zelensky concluded that the responsibility now lies on the international community, transcending the boundaries of "great powers," to address this looming spectre and forge a path toward sustainable peace and security.

Thus, experts from diverse fields convened, in the shadow of recent geopolitical upheavals, particularly the Ukrainian war. The goal is to dissect and understand the challenges and strategies to respond to potential radiation disasters effectively, navigating through complex themes, from geopolitical nuances to regional preparedness, the convergence of hybrid and nuclear threats, and the subtleties of effective communication in case of nuclear hazard.

The first theme that prominently emerged was the deep-seated interconnection between the geopolitical and strategic dimensions of medical response in radiation

N. Iancu (✉)
West University of Timisoara, Timişoara, Romania

Constanta Maritime University, Constanta, Romania
e-mail: niculae.iancu@e-uvt.ro; nicu.iancu@marcyscoe.org

F.-C. Cirstoiu et al., *Medical Response Strategy in Case of Radiation Emergency Caused by the War in Ukraine*, NATO Science for Peace and Security Series B: Physics and Biophysics, https://doi.org/10.1007/978-94-024-2266-5_15

disaster management. From a geopolitical lens, the complex interplay between nations, notably in the backdrop of the Ukrainian war, has highlighted that radiation emergencies are not solely domestic issues. They are blended with international politics, collective defence policies, and systemic power distribution and dynamics. The potential weaponisation of nuclear sites and energy can have a cascading impact on regional and global stability. Moreover, the shared vulnerabilities among the NATO Eastern Flank countries highlight the pressing need to foster collaborative efforts across borders under NATO guidance, emphasising synchronised initiatives and actions.

Strategically, the focus on medical response underscores the paramount importance of human security in the face of radiological threats. The delicate balance of ensuring immediate medical care while managing radiological concerns and the intricate communication dynamics between radiation protection professionals and healthcare providers are central to achieving beneficial outcomes. Various experts reiterated the need for an all-encompassing approach that marries medical expertise with technological advancements, robust infrastructure, and the active involvement of local communities. Hence, a significant worry is the potential burden on healthcare systems during a radiation emergency. This strain is not just logistical but also involves the knowledge dissemination and capability enhancement of frontline health workers who might be unversed in radiation-related medical emergencies. The need for tailored training programs, simulation exercises, and ongoing education was highlighted as paramount.

It is evident that the interwoven nature of geopolitics and strategic medical response in the case of radiation is not just a theoretical concept but a palpable reality faced by nations, especially those in conflict zones. In the wake of a potential radiological disaster, a nation's internal policies and preparedness measures cannot remain insulated from the larger international environment. Furthermore, the engagement of local communities cannot be stressed enough. They are the first line of response, often acting as the primary caregivers before official medical help can reach them. Their insights, traditional knowledge, and grassroots networks can prove invaluable in times of crisis, ensuring that aid goes where it's most needed and that communication remains transparent and efficient.

Second, the centrality of effective communication in the context of radiation disaster management emerged as a crucial theme. From a strategic standpoint, effective communication acts as the linchpin that binds together the myriad aspects of radiation disaster response. In the volatile environments created by events like the Ukrainian war, the narrative presented to the world holds significant power. Transparent, timely, and truthful communication can deter misinformation campaigns, prevent unnecessary panic, and ensure that both national and international stakeholders are aligned in their understanding and response to the situation.

Also, communication isn't merely about disseminating information. It demands a sophisticated grasp of its recipients, whether the affected populace, allied countries, or global organisations. Communication challenges are manifold in a digital era where misinformation can spread rapidly on digital platforms. Addressing these threats isn't just about debunking false narratives, but it also refers to building a foundation of trust and credibility in states and governments.

Furthermore, precise and efficient communication emphasises the dynamic between radiation protection professionals and healthcare providers. This intersection is where science meets care, and the synergy between the two can dramatically impact outcomes. A transparent and efficient flow of information ensures that both groups can function optimally, harnessing their expertise. By understanding and respecting the strengths and boundaries of each, they can ensure immediate relief and a sustainable and holistic approach to patient care in the aftermath of radiation emergencies.

Building on this premise, the knowledge of potential radiation hazards, exposure levels, and mitigation strategies is vital for radiation protection professionals. Conversely, healthcare providers bring to the table their expertise in diagnosing, treating, and providing long-term care for patients affected by radiation. When these two disciplines converge, it creates a formidable cooperation that enhances the efficacy of disaster response.

However, it's not without its challenges. The nuances of radiation science may be complex and beyond the purview of general medical practice. Similarly, the intricate medical needs and psychological ramifications of patients experiencing radiation exposure might be outside the immediate expertise of radiation professionals. This is where clear, continuous, and jargon-free communication becomes the bridge. Workshops and joint training sessions simulating real-life radiation crisis scenarios are potential avenues to foster this collaborative environment. By placing both professionals in a controlled setting, it allows them to understand the challenges, responsibilities, and capabilities of their counterparts. Such proactive initiatives can also lead to the development of standardised protocols and guidelines that can be swiftly deployed during real emergencies.

Moreover, with the evolution of technology and increasing integration of AI and telemedicine in healthcare, an untapped potential exists to refine this dynamic further. Digital platforms can facilitate real-time data exchange, provide diagnostic assistance, and even offer virtual training modules tailored for such interdisciplinary collaboration. Consequently, by harnessing the strengths of both fields and continuously investing in their collaborative potential, we can ensure that our response to radiation emergencies is not just immediate but also comprehensive, compassionate, and forward-thinking.

Third, is the theme of addressing the challenges and seizing opportunities in regional nuclear crisis preparedness. The workshop discussions underscored the sobering reality that the preparedness level of countries in the Black Sea region to handle nuclear crises varies, posing challenges as well as opportunities. The recent events surrounding Ukraine have brought into sharp focus the critical need for a robust and coordinated regional response mechanism. While some states, due to historical or strategic reasons, have invested considerably in nuclear disaster preparedness and response infrastructure, others still have significant gaps in terms of resources, training, and strategy.

Proactive investments in strengthening regional preparedness mitigate the risks of nuclear incidents and serve as a confidence-building measure among neighbouring countries. Collaborative regional training programs, joint simulation exercises, sharing best practices, and creating a centralised repository of resources and expertise were some common suggestions. However, with varying economic capacities among the countries in the Black Sea region, there's a pressing need to establish mechanisms for equitable distribution of resources. International organisations, non-profits, and philanthropic entities could play a pivotal role in bridging these resource gaps. By providing financial support, technology transfers, and expertise, they can ensure that all countries are adequately equipped to tackle nuclear crises, irrespective of their economic stature. Education and public awareness are other key societal resilience facets from the regional perspective. Educating communities about the risks of nuclear incidents and the necessary preventive and responsive measures is crucial. By fostering a culture of awareness, individuals become proactive stakeholders in their safety. Furthermore, the need for research and innovation in the realm of nuclear safety was highlighted. Joint research initiatives, technology-sharing agreements, and innovation hubs dedicated to nuclear safety can be instrumental in staying ahead of potential threats.

However, despite the evident challenges, there was also a sense of optimism. The regional cooperation exhibited during recent crises and the active involvement of international bodies like NATO provide a strong foundation. The value of diplomatic dialogues cannot be underestimated. Bilateral and multilateral talks among the Black Sea countries and beyond can pave the way for transparency, trust, and collaborative action. Regular dialogues can prevent misunderstandings, ensure shared intelligence, and facilitate swift, coordinated emergency responses. Building on this, the countries in the region have a unique opportunity to lead by example, demonstrating that collective preparedness can act as a deterrent to potential nuclear threats and ensure the safety and well-being of their citizens.

Also, hybrid threats, particularly in the context of the Russian-Ukrainian war, have become increasingly sophisticated and encompass a broader spectrum of tactics. The adaptation to these hybrid threats necessitates a multifaceted approach. Beyond just counteracting the immediate threats, nations need to invest in long-term solutions like strengthening their digital infrastructure, promoting media literacy among citizens, and fostering solid international collaborations. Engaging in partnerships with global tech companies and platforms can also be a way to keep a check on the spread of misinformation and devise mechanisms to debunk it instantly.

The Republic of Moldova is a critical example of the intensifying attempts to destabilise nations using a mix of conventional and unconventional methods—ranging from energy and political pressures to information warfare. Russia's strategic leveraging of hybrid tactics, especially in the digital information space, underscores the vulnerabilities of nations with significant Russian-speaking populations. Disinformation campaigns, amplified by the reach and speed of social media platforms, have targeted the immediate populace and aimed at influencing perceptions on a global scale. Such narratives, often intertwined with religious or conspiratorial themes, spread discord, challenge democratic values, and threaten the societal cohesion of targeted nations.

Romania's assessment of potential scenarios further illustrates the importance of recognising the varied forms these threats can manifest in. In this context, preparedness extends beyond just military might or political alliances. It encompasses resilience against misinformation, infrastructural vulnerabilities, and potential societal fractures. The situation in Moldova and the experiences of Romania underscore the need for nations to be proactive rather than reactive. Comprehensive intelligence gathering and sharing mechanisms, coupled with establishing rapid response teams, can act as first lines of defence against these ever-evolving threats.

The emphasis on the role of Kremlin proxies, as demonstrated in Moldova's case, and the challenges they pose to stability show that hybrid threats are not just about overt actions but also covert influence campaigns. The ever-evolving nature of these threats necessitates continuous evaluation and adaptation of defence and response strategies, ensuring nations stay one step ahead in this complex game of geopolitical chess. So, while the landscape of hybrid threats continues to change and evolve, with collective effort, strategic planning, and international cooperation, democratic nations can stand resilient against hybrid threats, ensuring the safety and stability of their regions and the broader global community.

Moreover, the convergence of hybrid and nuclear threats can no longer be viewed in isolation, and their overlap introduces a level of unpredictability that needs to be addressed proactively. Integrating hybrid strategies into the nuclear domain marks a concerning evolution in global security dynamics. The circumstances of the Ukrainian war showcase how potential nuclear crises can be exacerbated by, or even intertwined with hybrid tactics.

Nuclear geopolitics, already a complex web of power dynamics, strategic deterrence, and mutual assurance, is now further complicated by the infusion of misinformation, disinformation, and digital warfare. The weaponisation of nuclear narratives, as seen in the dissemination of news related to the use of nuclear weapons in Ukraine, amplifies the risks and uncertainties surrounding actual nuclear events. With its rapid spread of information, the digital sphere can wave the flames of a localised incident into a regional or global crisis, complicating timely and appropriate responses.

Furthermore, as Russia seeks to redefine its sphere of influence and geopolitical standing, the fusion of hybrid threats with nuclear posturing poses significant challenges for countries from the NATO Eastern Flank and NATO partners in the region. Nations must bolster their capabilities to discern truth from falsehood, ensuring that their responses are based on factual assessments and not manipulated narratives. Collaborative efforts within the region and with international allies are essential to build resilience against this dual menace of hybrid nuclear threats. As evident from the discussions, nations must enhance their technological infrastructure to detect and combat the multifaceted challenges presented by the digital age. This includes developing advanced cybersecurity measures, investing in AI-driven intelligence analysis, and fostering collaboration in cyber forensics. Regular dialogues, joint military exercises, and intelligence sharing can help anticipate potential challenges and formulate cohesive strategies to address them. Thus, as the threats amalgamate and become increasingly complex, the need for unity, foresight, and innovation becomes ever more crucial. The international community must unite to forge a path of security and stability in this challenging era.

Lastly, the prominent role of significant international organisations grew into a central theme. The perspectives and functions of major international organisations have emerged as critical pillars in the discussions surrounding the multifaceted challenges of radiation emergencies, hybrid threats, and regional geopolitical dynamics. Organisations like NATO and the EU are pivotal in shaping regional preparedness, response strategies, and policy frameworks. Their commitment to fostering collaboration, ensuring collective defence, and supporting capacity-building efforts underscores the importance of multilateralism in addressing contemporary security challenges.

More specifically, the EU's initiatives, such as the civilian partnership mission EUPM Moldova, spotlight the organisation's proactive approach to enhancing security sector resilience and its commitment to an integrated regional response. Similarly, NATO's influence on fostering transnational collaborations emphasises the alliance's commitment to collective defence and mutual support.

Furthermore, the insights and expertise these organisations provide serve as a bridge to align national strategies with international best practices, standards, and expectations. The continuous dialogue with and among these organisations offers nations in the region a platform to share experiences, lessons learned, and innovative approaches to deal with shared vulnerabilities and threats.

Therefore, the perspectives and contributions of major international organisations are indispensable. Their ability to harmonise efforts, share resources, and establish common frameworks offers a pathway for nations in the region to navigate the intricate geopolitical landscape effectively, enhance their preparedness, and collectively address the myriad challenges of the twenty-first century.

The emphasis on preparedness, collaboration, and communication resonated deeply, spotlighting the high stakes in radiation incidents and hybrid threats. The discussions underscored the urgent need for a unified international response, with entities like NATO leading the way.

The ever-changing geopolitical landscape and the rise of new threats highlight the indispensable strength found in unity, collective wisdom, and a global commitment to safety. Policymakers, researchers, and experts alike are now tasked with transforming these insights into actionable steps. While the workshop may have concluded, the shared dedication to enhancing cooperation and fortifying resilience remains. Together, we move forward, better equipped to tackle the radiological challenges of the future.

The End of (Nuclear) History Revisited: Could Russia Break the Nuclear Taboo?

Dan Dungaciu and Ioana Constantin Bercean

Abstract The Cold War storytellers does not often refer to that era as of a period of superpowers cooperation, but like a nuclear competition. And actually, they are not entirely wrong. Shortly after the end of the Second World War, the nuclear race between the United States and the Soviet Union became the paramount of the Cold War era. The invention of nuclear weapons ushered in a revolution in international politics and particularly in the way that states think about and plan for war.

Since launching its full-scale invasion of Ukraine, Russia has conducted an intense campaign of nuclear signaling and threats designed to provide it with a strategic advantage. Whatever Russia's nuclear strategy is, Russian officials have made many statements about nuclear weapons that appear to go beyond the published doctrine, threatening to potentially use them in situations that do not meet the conditions described.

In this context, the question arises how worried should we be? Through nuclear insinuations, the Kremlin appears to reinforce three goals: drawing a clear red line against NATO boots on the ground in Ukraine, to climb the escalation ladder so, when the tipping point is reached, the West might accept the idea of some negotiations. The dynamic behind escalation is provided by either the tit-for-tat mentality and the bizarre game of nuclear look-at-me, which means that if one side ups the ante, the other is sure to follow, or the kind of miscalculation or chance occurrence common in war.

So far, the 79-year-old taboo against the use of nuclear weapons has held, but the world cannot take it for granted. To preserve and strengthen the consensus against

D. Dungaciu (✉)
Department of Sociology, University of Bucharest, Bucharest, Romania

International Department, Institute of Political Science and International Relations, Bucharest, Romania
e-mail: dan.dungaciu@unibuc.ro

I. C. Bercean
Department of Sociology, University of Bucharest, Bucharest, Romania

F.-C. Cirstoiu et al., *Medical Response Strategy in Case of Radiation Emergency Caused by the War in Ukraine*, NATO Science for Peace and Security Series B: Physics and Biophysics, https://doi.org/10.1007/978-94-024-2266-5_16

nuclear weapons use and threats of use, we must sustain pressure against those who might try to break the nuclear taboo.

> *When he woke in the woods in the dark and the cold of the night he'd reach out to touch the child sleeping beside him. Nights dark beyond darkness the days more gray each one than what had gone before. Like the onset of some cold glaucoma dimming away the world. His hand rose and fell softly with each precious breath. He pushed away the plastic tarpaulin and raised himself in the stinking robes and blankets and looked toward the east for any light but there was none.* Cormac McCarthy, *The Road* (2006)

1 Dark Beyond Darkness

The Cold War storytellers does not often refer to that era as of a period of superpowers cooperation, but like a nuclear competition. And actually, they are not entirely wrong. Shortly after the end of the Second World War, the nuclear race between the United States and the Soviet Union became the paramount of the Cold War era. And by 1987, the nuclear stockpile of the USSR reached its maximum at 40,723 weapons, while the USA had 23,490 [22]: 17; *see also* [34]: 11; [36]: 17; [5].

The invention of nuclear weapons ushered in a revolution in international politics and particularly in the way that states think about and plan for war. As Robert Jervis put it, nuclear weapons "changed the way people think and nations behave" [15]: 9. These new types of capabilities were transformational weapons and entered the international scene with four proposals: the bombs' effects were devastating; cities were their natural targets; there was no obvious defense; and they could win wars [7]: vii. Over the years each of those propositions were either challenged, or became qualified, yet the first impression counts and ever since the first, and only, use of the nuclear weapons, scholars and politicians alike have puzzled over their effects on international and domestic affairs. Yet, there is one sure thing: the detonation of just one of the approximately 13,355 warheads ready-to-use that exist today would cause unimaginable destruction to any large modern city, and even a very limited nuclear exchange could risk the annihilation of entire societies [9]: 1. Indeed, some scientists suggest that the use of nuclear weapons could lead to a nuclear winter where the world might suffer years of freezing temperatures due to the radioactive fallout released into the atmosphere, and where all life forms would struggle to survive [31]: 257. Others suggest that a nuclear war could cause human society to return to the Stone Age, or even end the human race entirely. Or as Joseph Nye put it,

> The prospect of a nuclear war is horrifying. It brings us face to face not only with death, but with destruction of the civilization that makes our life meaningful. It might even destroy our species. There is no precedent for the challenge that nuclear weapons present to our physical and moral lives. [26]: ix

The dream of a world without nuclear weapons has never faded since the dawn of the nuclear age, and in September 2017, a group of states signed the landmark *Treaty on the Prohibition of Nuclear Weapons*, also known as the *Nuclear Ban Treaty*. As a matter of fact, former US President Barack Obama, has started his first term in office by setting a new direction in US nuclear weapons policy and articulated the benchmarks of this direction during his monumental speech in Prague on April 5, 2009 [36]: 5. However, while the *Ban Treaty* may be symbolic of a renewed desire for movement towards a nuclear-free world, none of the current nuclear-armed states have signed up to it. Moreover, all nuclear-armed states are currently working to replace or modernize their existing nuclear weapons systems and support infrastructure, in a context where the *Nonproliferation Treaty* (NPT) is the last standing guardianship of the nuclear security assurances.

Nevertheless, the Cold War's demise saw a brief but relatively industrious phase of nuclear disarmament and threat reduction action, yet the global level of nuclear weapons still remained high and, and as Melissa Gillis argued, "the number of nuclear weapons has been significantly reduced, yet they continue not only *to exist*, but also *to be central to the security doctrines of those States that possess them*" [10]. And indeed, after 1991, most of the observers were sure that while the threat of an all-out nuclear war has diminished, the likelihood that nuclear weapons might be used in some way was increasing. This assessment was widely shared, and has become almost axiomatic among policymakers and within the strategic studies community. Yet, little systematic thought has been given to what the future use of nuclear weapons might look like. Therefore, decision-makers are often confronted by "precarious ignorance" [35]: 27 of inadequate *information-cum-potentially* disastrous consequences of the decision made on that narrow knowledge base. Even though an impressive number of scholars have made rivers of ink flow about the danger of nuclear weapons, in order to be taken seriously by policymakers, they have to learn to speak their language, to get and stay inside the dominant discourse. Many serious social scientists in general are reluctant to simplify their analyses and conclusions, airbrush their caveats and qualifications, and present recommendations with the degree of certainty that decision-makers demand. And so, they end up sharing the fate of Cassandra: blessed with the gift of prophecy but cursed with not being believed (*Ibid.*)

Since the beginning of the twenty-first century, Paul Bracken, Andrew Futter, Jeffrey Larsen, Lawrence Freedman, Jeffrey Michaels or Viping Narang, among others, have warned that "atomic weapons have returned for a second act" [2]: 7; *see also* [20]: 22; [25]: 11; [7]: 579 and yet, the policymakers were still prisoners of the rhetoric of the 1990s, that with the end of the Cold War the chapter of nuclear danger became history.

This article aims to address exactly that shortfall.

2 The End of (Nuclear)History Revisited

One that has been actually cancelled by the February 24, 2022 Russian invasion in Ukraine, followed by the implicit or explicit threats of President Vladimir Putin or of his political and media pundits regarding the use of nuclear weapons. At the time, the international community took a major blow. Already hit by 2 years of the COVID-19 pandemic and deteriorating interstate relations, it stood in horror as Russian forces unleashed an unprovoked war on a neighboring country [11]: 6. The significance of the Russian invasion is even more acute if analyzed through the lens of nuclear weapons. Europe and the world are now held hostage by a Russian president willing to resort to the threat of nuclear weapons use. And so, the nuclear weapons re-entered the center stage, with the alarming possibility of their use in a manner that had mostly been forgotten. Although the actual fighting is taking place between nuclear-armed Russia and non-nuclear Ukraine, the threatening shadow of the nuclear weapons possessed by the United States and NATO is palpable [32]: 12.

Since the Cuban Missile Crisis in 1962, this is the first real engagement "between the US and Russia where they are indirectly yet directly involved" (*Ibid.*). On Black Saturday—October 27, 1962—leaders in the US, USSR, and Cuba came close to blowing up the world [1]: xiii. Then the world escaped, in good measure, because of luck. How and where will it get this time?—it is a question to which no one dares (or knows how) to offer an answer. We do not know what exactly "winning" means for Putin and how far he is willing to go to achieve his insistent goal of "denazifying" Ukraine [4]. Just as no one has yet offered a clear strategy by which a nuclear power can be defeated without starting an all-out war.

Since launching its full-scale invasion of Ukraine, Russia has conducted an intense campaign of nuclear signaling and threats designed to provide it with a strategic advantage. Several senior Russian politicians, including President Vladimir Putin, former president and prime minister Dmitry Medvedev, and foreign minister Sergey Lavrov, have made a number of statements widely seen as threatening the use of nuclear weapons [13].

The Kremlin aims to not only deter the West from directly interfering in its attempts to conquer Ukraine but also to limit Western political, economic, and military support for the country. Moscow has been particularly successful in this: not only has NATO refrained from direct intervention—such as by setting up a no-fly zone—but Western nations have avoided forms of military assistance for Ukraine that fall within their rights as third parties in the conflict.

The famous article of Sergei A. Karaganov is emblematic in this respect [17]. Karaganov is a prominent political scientist and foreign policy expert who has advised both Boris Yeltsin and Vladimir Putin. In a controversial essay published in June 2023, he argued, in a very "Russian" style, that Russian nuclear preemption was needed to reawaken NATO fears of nuclear attack and reestablish deterrence, to prevent an otherwise inevitable escalation of the war in Ukraine to global thermonuclear war: „The creation of nuclear weapons was the result of divine intervention. Horrified to see that people, Europeans and the Japanese who had joined them, had unleashed two world wars within the life-span of one generation, sacrificing tens of millions of lives, God handed a weapon of Armageddon to humanity to remind those who had lost the fear of hell that it existed. It was this fear that ensured relative peace for the last three quarters of a century. That fear is gone now. What is happening now is unthinkable in accordance with previous ideas about nuclear deterrence: in a fit of desperate rage, the ruling circles of a group of countries have unleashed a full-scale war in the underbelly of a nuclear superpower. That fear needs to be revived. Otherwise, humanity is doomed. What is being decided on the battlefields in Ukraine is not only, and not so much, what Russia and the future world order will look like, but mainly whether there will be any world at all or the planet will turn into radioactive ruins poisoning the remains of humanity. By breaking the West's will to continue the aggression, we will not only save ourselves and finally free the world from the five-century-long Western yoke, but we will also save humanity. By pushing the West towards a catharsis and thus its elites towards abandoning their striving for hegemony, we will force them to back down before a global catastrophe occurs, thus avoiding it. Humanity will get a new chance for development" [17].

Importantly, we are not arguing that Karaganov is necessarily an influential adviser to Putin these days. Karaganov's role is different. He exaggerates, hyperbolizes, leads to extremes of certain polices or potentialities. He is the "Russian" on duty, as it appears in the Western mind, a consequence of Russian literature or mythology, especially in the vein of Dostoievsky's novels: demonic, capable of anything, irrational, ready to sacrifice anything for an idea, no matter how crazy it would be, with a different kind of reason than the Western one, and therefore impossible to fully understand. Karaganov is not right in the pure sense of the term, but he conveys the message that even Western assessments are not necessarily right to the end.

Nevertheless, it is a considerable worry that these sorts of outrageous nuclear threats are being traded frequently in Moscow.

The possibility of Russia using tactical nuclear weapons, and the risk of broader nuclear escalation, has been widely discussed by commentators and in the media. And, as with any war scenario, the pundits' voices have split into two schools of thought. The first camp is that of the optimists, who believe that the risk of a nuclear war is low (but not impossible!). Their reasoning is simple: the countries that are building up and brandishing their nuclear capabilities are bluffing [21]: 46. Nuclear weapons cannot cover conventional military weakness because threats to escalate, even by a desperate enemy, are not credible. According to these soothing voices, giving credence to the nuclear bluster of weak enemies is misguided and plays squarely into their hands (*Ibid.*). And yet, is there a real possibility that Putin would order a nuclear weapon to be dropped on Ukraine? If it were not on the table, we would not even be considering it. That's what the pessimists think too. For them, the risk of nuclear escalation during conventional war is much greater than is generally appreciated (*Ibid.*). The conundrum that US adversaries face today—how to convincingly threaten escalation and bring a nuclear-armed opponent to a stalemate—was solved decades ago by the US and its NATO allies. Back then, the West developed a strategy of coercive nuclear escalation to convince the USSR that NATO allies would actually use nuclear weapons if they were invaded. Today, US rivals have adopted NATO's old nuclear strategy (*Ibid.*) and developed their own options for credible escalation. It would be tragic for Washington (and for the rest of the world) to stumble into nuclear war because it discounted the very strategy that it invented decades ago.

These are difficult questions to which everyone would like some firm answers. But it will take a long time (if ever), and many analyses, until the light is shed on them. For now, we can try to decipher what the nuclear arsenal means in Russian politics and religion, in an attempt to understand if behind Moscow's threats there is something more than a simple bluff rhetoric.

3 Nuclear Weapons in Russian Politics and Strategy

Just 2 years before invading Ukraine, Russia was in the midst of modernizing its nuclear arsenal, to replace Soviet-era weapons with newer systems. In December 2020, President Vladimir Putin reported that modern weapons and equipment now make up 86% of Russia's nuclear triad [29], compared to the previous year's 82% [28]. He additionally noted that he expects that number to rise to 88.3% in 2021. As in previous years, Putin's remarks emphasized the need for Russia's nuclear forces to keep pace with Russia's competitors:

> It is absolutely unacceptable to stand idle. The pace of change in all areas that are critical for the Armed Forces is unusually fast today. It is not even Formula 1 fast—it is supersonic fast. You stop for one second and you start falling behind immediately. [29]

Russia's nuclear modernization program was motivated in part by Moscow's strong desire to maintain overall parity with the US, but also by the Russian leadership's apparent conviction that the US ballistic missile defense system constitutes a real future risk to the credibility of Russia's retaliatory capability. Policy and strategy aside, the development of multiple weapon systems also indicates the strong influence of the military-industrial complex on Russia's nuclear posture planning. However, one can wonder if this strategy of renewing the nuclear arsenal was made in the perspective of what would follow in February 2022. In any case, there is certain that Russia is now living through a self-proclaimed "nuclear renaissance" [16]: 6. If the atom fell on hard times in the 1990s under Boris Yeltsin, then it has found full support and full funding under President Vladimir Putin. The economic and political crises of the 1990s that saw many industries falter, and scientific research cease in many institutes, have passed. The economy has rebounded on the sales of oil and gas (*Ibid.*), and Putin has determined to build a strong state on resource extraction, and to recreate Russia as a scientific superpower through achievements in space and nuclear power.

The international debate about Russia's nuclear strategy has reached a new level of intensity, particularly after the Trump administration published its Nuclear Posture Review in February 2018. The Nuclear Posture Review claims that,

> Russian strategy and doctrine emphasize the potential coercive and military uses of nuclear weapons. It mistakenly assesses that the threat of nuclear escalation or actual first use of nuclear weapons would serve to 'de-escalate' a conflict on terms favorable to Russia. (US Defense Department, 2018: 8)

Specifically, the document claims, "Moscow threatens and exercises limited nuclear first use, suggesting a mistaken expectation that coercive nuclear threats or limited first use could paralyze the United States and NATO and thereby end a conflict on terms favorable to Russia." This so-called "escalate to de-escalate" [19]: 92 doctrine "follows from Moscow's mistaken assumption of Western capitulation on terms favorable to Moscow" (US Defense Department, 2018: 30).

John Hyten (gen. r.), the former head of the US Strategic Command, reacted to "Russia's destabilizing doctrine on what some call escalate to deescalate" by saying: "I really hate that discussion. I've looked at the Russian doctrine. I've looked at Russian writings. It's not escalate to deescalate, it's escalate to win. Everybody needs to understand that" [14]. Some have suggested that Russian leaders are signaling a willingness to use nuclear weapons even before an adversary retaliates against a Russian conventional attack by "employing the threat of selective and limited use of nuclear weapons to forestall opposition to potential aggression" [23]. The implication is that Russia would potentially use nuclear weapons first to scare an adversary into not even defending itself [19]: 93.

Whatever Russia's nuclear strategy is, Russian officials have made many statements about nuclear weapons that appear to go beyond the published doctrine, threatening to potentially use them in situations that do not meet the conditions described. For example, officials explicitly threatened to use nuclear weapons against ballistic missile defense facilities, and in regional scenarios that do not threaten Russia's survival or involve attacks with weapons of mass destruction [37]. However, policy and strategy aside, the development of multiple weapon systems also indicates the strong influence of the military-industrial complex on Moscow's nuclear posture planning, and they are also a sign of Russia's readiness to use its nuclear capabilities.

4 The Russian Tactical Nuclear Weapons Posture

Discussions about a possible use by Russia of a 'tactical nuclear weapon' (TNW) dominated the analytical environment shortly after the start of the war in Ukraine. However, there is no simple, clear, and universally valid definition of TNWs [24]: 27. This term is often used interchangeably with various synonyms, such as nonstrategic, sub-strategic, short-range, shorter-range, theater and battlefield weapons. Although 'tactical' is an imprecise and not completely satisfactory designation, this term is nonetheless employed throughout all the public discourses [6]: 4. Nevertheless, conceptual clarity is indispensable for fully understanding the terms of the ongoing debate, because unsatisfactory definitions not only confuse theoretical discussions regarding the Russian posture on TNWs, but it might mislead the public opinion.

Two main features determine the nature of TNWs: first, the concept of a tactical weapon relates to its short range and second, the term 'tactical' also refers to the function of a weapon. TNWs are conceptually designed to perform specific military functions that involve supporting conventional military operations on the battlefield (*Ibid.*: 5). In other words, TNWs are intended for use in combat to achieve circumscribed objectives in a nuclear war that is envisaged as limited and non-suicidal. Their purpose is to win a single battle rather than to immediately decide the outcome of an entire war. Although this element of the 'tactical' classification adds a degree of definitional clarity to this term (*Ibid.*), the notion of a tactical mission can be controversial, as disputes may arise regarding whether a particular target is *tactical* or *strategic* in nature. Moreover, no certainty exists about the actual *strategic effects* and *political consequences* that the use of tactical nuclear weapons may produce on an adversary [3].

As for Russian TNWs, they remain a contentious issue in US—Russian relations, in NATO—Russian relations, and in transatlantic and intra-European relations. This discord stems primarily from the significant imbalance in the number of weapons possessed by Russia and the US; estimates suggest that there are 2000 Russian TNWs while the us has only 500–600 [33]: 158. The US and Russia do not exchange data on the numbers and locations of their TNWs; as a result, the publicly available data that exist are unofficial expert estimates. Over more than a decade, Russia's main condition for starting negotiations on TNWs has been the withdrawal of the remaining US tactical nuclear weapons from Europe (*Ibid.*: 158–159).

However, more important than sheer numbers are the Russian will to use those capabilities or not. In a 2007 speech, the Chief of the 12th GUMO (the main directorate of the Russian Minstry of Defense), General Vladimir Verkhovtsev said that Russia no longer deployed TNWs on surface ships and submarines but "if necessary... it could... [and] no one should doubt that" (*Ibid.*: 160). Although this statement is over a decade old, its echoes can be found in the current confrontation between Russia and Ukraine. In October 2023, President Vladimir Putin suggested that Moscow might restart nuclear testing and just earlier that that, Russia deployed some TNWs to the neighboring and friendly country, Belarus.

5 The Nuclear Taboo

In this context, the question arises how worried should we be? A bit perhaps, but not too much, Lawrence Freedman believes. As he has noted in April 2024, the actual use of non-strategic nuclear weapons in Ukraine would be inconsistent with Russian nuclear doctrine and with what Vladimir Putin himself has said. Russia's 2020 nuclear policy statement provided that the Kremlin could use nuclear weapons in the event of "aggression against the Russian Federation with the use of conventional weapons when the very existence of the state is in jeopardy" [8]. No serious military analyst believes that Russia's existence is imperiled. He goes further and points out that when Putin raises nuclear weapons, he directs the threat against the US and its allies that are supporting Ukraine, not against Ukraine itself (*Ibid.*). Also, through nuclear insinuations, the Kremlin appears to reinforce three goals: drawing a clear red line against NATO boots on the ground in Ukraine, to climb the escalation ladder so, when the tipping point is reached, the West might accept the idea of some negotiations. However, once he has stepped on the escalator, Putin is bound to take it to the next level, even if he will soon conclude that he would rather not do it. The dynamic behind escalation is provided by either the *tit-for-tat* mentality and the bizarre game of nuclear *look-at-me* [12], which means that if one side ups the ante, the other is sure to follow, or the kind of miscalculation or chance occurrence common in war.

Russian officials also must consider that the use of a nuclear weapon, against any target, would open a Pandora's box [27] and no one really can foresee its consequences. A nuclear weapon has not been used in anger since 1945—and then it was used by the US, then the only country with nuclear arms, not in a world of multiple nuclear weapons states. The risks and potential costs of such an action for Moscow would be immense. Still, the US and its allies cannot gamble the world's future on Vladimir Putin's mood or on Russian officials to stop him from using nuclear weapons. Nuclear brinkmanship is a dangerous game. The US also should explore the possibility of a modest trilateral Chinese-Russian-US dialogue [30]: 6 on nuclear risk reduction and all states concerned about nuclear escalation also could urge the five NPT nuclear-weapon states to express support for the 1973 *Soviet-US Agreement on the Prevention of Nuclear War* [18]: 3, which requires that states refrain from nuclear threats and, in times of increased risk of nuclear conflict, "immediately enter into urgent consultations with each other and make every effort to avert this risk" (*Ibid*).

So far, the 79-year-old taboo against the use of nuclear weapons has held, but the world cannot take it for granted. Tragically, the end of Russia's war on Ukraine is nowhere in sight. To preserve and strengthen the consensus against nuclear weapons use and threats of use, we must sustain pressure against those who might try to break the nuclear taboo.

References

1. Blight JG, Lang JM (2018) Dark beyond darkness: the Cuban missile crisis as history, warning, and catalyst. Rowman & Littlefield, London
2. Bracken P (2012) The second nuclear age. Strategy, danger, and the new power politics. St. Martins's Press, New York
3. Cappello JT, Hall GM, Lambert SP (2002) Tactical nuclear weapons: debunking the mythology. INSS Occasional Paper, No. 46, United States Air Force Institute for National Security Studies
4. Chinchilla M (2022) The Russian nuclear threat: the weight of uncertainty. Global Affairs and Strategic Studies. https://en.unav.edu/web/global-affairs/la-amenaza-nuclear-rusa-el-peso-de-la-incertidumbre. Accessed 4 June 2024
5. Davenport K (2024) Nuclear weapons: who has what at a glance. Arms Control Association. https://www.armscontrol.org/factsheets/nuclear-weapons-who-has-what-glance?check_logged_in=1. Accessed 14 May 2024

6. Foradori P (2013) Introduction: debating the last remaining case of the forward deployment of nuclear weapons. In: Foradori P (ed) Tactical nuclear weapons and Euro-Atlantic security: the future of NATO. Routledge, New York, pp 1–26
7. Freedman L, Michaels J (2019) The evolution of nuclear strategy, 4th edn. Palgrave Macmillan, London
8. Freedman L (2024) Escalation, red lines, risk and the Russo-Ukraine war. https://samf.substack.com/p/escalation-red-lines-risk-and-the. Accessed 25 May 2024
9. Futter A (2021) The politics of nuclear weapons, 2nd edn. Palgrave Macmillan, Cham
10. Gillis M (2009) Disarmament: a basic guide. United Nations Office for Disarmament Affairs United Nations, New York. www.un.org/disarmament/HomePage/ODAPublications/AdhocPublications/PDF/guide.pdf. Accessed 16 May 2024
11. Giovannini F (2022) Negative security assurances after Russia's invasion of Ukraine. Arms Control Association 52(6):6–11
12. Gottemoeller R (2024) The changing nuclear mind game. Foreign Policy. https://foreignpolicy.com/2024/05/15/nuclear-weapons-mind-game-deterrence-escalation-putin-russia-threats-ukraine-war-biden/. Accessed 24 June 2024
13. Gressel G (2022, July 7). Shadow of the bomb: Russia's nuclear threats – European Council on Foreign Relations. ECFR. https://ecfr.eu/article/shadow-of-the-bomb-russias-nuclear-threats/. Accessed 24 June 2024
14. Hyten JE (2017) 2017 deterrence symposium closing remarks, Omaha, NE. http://www.stratcom.mil/Media/Speeches/Article/1264664/2017-deterrence-symposium-closing-remarks/. Accessed 22 May 2024
15. Jervis R (1989) The meaning of the nuclear revolution: statecraft and the prospect of Armageddon. Cornell University Press, Ithaca
16. Josephson PR (2022) Nuclear Russia. The atom in Russian politics and culture. Bloomsbury, London
17. Karaganov SA (2023) A difficult but necessary decision. Russia in Global Affaires 13.06. https://eng.globalaffairs.ru/articles/a-difficult-but-necessary-decision/. Accessed 17 May 2024
18. Kimball DG (2024) Russua's war and the specter of nuclear conflict. Arms Control Association 54(5):3
19. Kristensen HM, Korda M (2021) Russian nuclear weapons, 2021. Bull At Sci 77(2):90–108. https://doi.org/10.1080/00963402.2021.1885869
20. Larsen J, Kerry K (2014) On limited nuclear war in the 21st century. Stanford University Press, Stanford
21. Lieber KA, Press DG (2024) The return of nuclear escalation. Foreign Aff 102(6):45–55
22. Light M (2003) 100 Suns: 1945–1962. Jonathan Cape, London
23. Miller F (2015) Keynote Speech at the 2015. USSTRATCOM deterrence symposium. https://www.stratcom.mil/Media/Speeches/Article/986412/keynote-2015-usstratcom-deterrence-symposium/. Accessed 22 May 2024
24. Müller H, Schaper A (2000) Definitions, types, missions, risks and options for control: a European perspective. In: Potter WC et al (eds) Tactical nuclear weapons options for control. United Nations Institute for Disarmament Research, New York, pp 19–51
25. Narang V (2014) Nuclear strategy in the modern era: regional powers and international conflict. Stanford University Press, Stanford, CA
26. Nye J (1988) Nuclear ethics. Collier Macmillan Publishers, London
27. Pifer S (2024) Holding one's nerve in the face of Russian nuclear threats. Brookings Institution. https://www.brookings.edu/articles/holding-ones-nerve-in-the-face-of-russian-nuclear-threats/. Accessed 24 June 2024
28. Russian Federation Defense Ministry (2019) Chief of the General Staff of the Russian Armed Forces – First Deputy Minister of Defence General of the Army Valery Gerasimov Meets with Representatives of the Military Diplomatic Corps Accredited in Russia. Press Release. http://eng.mil.ru/en/news_page/country/more.htm?id=12267331. Accessed 22 May 2024

29. Russian Federation Defense Ministry (2020) Installation of the Avangard Intercontinental Ballistic Missile in a Silo Launcher. https://youtu.be/64C8YBXU1SQ. Accessed 22 May 2024
30. Rusten L, Melamed M (2023) The three-competitor future: U.S. arms control with Russia and China. Arms Control Association 53(2):6–13
31. Sagan C (1983) Nuclear winter and climatic catastrophe: some policy implications. Foreign Aff 62(2):257–292
32. Sethi M (2022) Nuclear overtones in the Russia-Ukraine war. Arms Control Association 52(5): 12–14
33. Sokov NN (2013) Russia. In: Tactical nuclear weapons and Euro-Atlantic security: the future of NATO. Routledge, New York, pp 158–176
34. Sprod L (2012) Nuclear futurism: the work of art in the age of remainderless destruction. Zero Books, Washington
35. Thakur R (2017) Nuclear weapons and international security. Routledge, New York
36. Warren A (2014) The Obama administration's nuclear weapon strategy: the promises of Prague. Routledge, New York
37. Warren A (2015) Russia delivers nuclear threat to Denmark. The Local. http://www.thelocal.dk/20150321/russia-threatens-denmark-with-nuclear-attack. Accessed 22 May 2024

Russia's Hybrid Nuclear Compellence: Intentions and Consequences

Valeriia Hesse

1 Introduction

Intimidation, or the strategic use of fear to secure desired outcomes, is prominently featured in Russia's war against Ukraine in numerous ways: horrific acts of violence against civilians, widespread missile and drone strikes, the deliberate targeting of energy infrastructure, and nuclear threats. These tactics are designed to force Ukraine into submission while also discouraging external parties from providing support.[1] Two aspects of this tactics leverage nuclear fears: threats involving the use of nuclear weapons and the manipulation of radiological emergency risks, both of which exploit public fear of radiation.

2 Intentions

When examining Russia's coercive strategies, it is vital to consider the framework of compellence, which falls under the larger concept of coercion. Coercion involves using implicit violence to alter the target's perception of costs, either to prevent them

[1] Victoria Namestnik, "How Russian Propaganda Uses Intimidation Tactics," Detector Media, February 27, 2023, https://disinfo.detector.media/post/yak-rosiiska-propahanda-vykorystovuie-taktyku-zaliakuvannia; Olha Kotiv. "'There Is No Chance of Survival': How Ukraine Met Spring, and Putin's Winter Terror Failed," 24 Channel, March 1, 2023, https://24tv.ua/zalyakuvannya-ukrayini-zimoyu-plan-putina-zamoroziti-ukrayintsiv_n2265364; Claire Mills, Russia's use of Nuclear *Threats* during the Ukraine Conflict, (London, UK: House of Commons Library, 2023), p.12, https://researchbriefings.files.parliament.uk/documents/CBP-9825/CBP-9825.pdf

V. Hesse (✉)
Odesa Center for Nonproliferation (OdCNP), Odesa, Ukraine
e-mail: valeriia.hesse@gmail.com

F.-C. Cirstoiu et al., *Medical Response Strategy in Case of Radiation Emergency Caused by the War in Ukraine*, NATO Science for Peace and Security Series B: Physics and Biophysics, https://doi.org/10.1007/978-94-024-2266-5_17

from taking an action (deterrence, maintaining the status quo) or to compel them to take an action (compellence, altering the status quo). Coercion can thus involve both forcing action and restraining it. The main objective is to convince the adversary that the potential losses from acting in the first scenario, or from not acting in the second scenario, will far outweigh any possible gains.[2] A crucial aspect of this strategy is how the opponent perceives these threats.

Russia has employed nuclear weapons rhetoric and signaling on numerous occasions. The comprehensive list is available through the Center for Strategic and International Studies's "Nuclear Signaling during the War in Ukraine" database.[3] However, two particularly significant instances stand out:

- Russian President Vladimir Putin made a statement that was largely seen as his initial nuclear threat on February 24, 2022. He cautioned that any interference by other nations in Russia's actions in Ukraine would result in an immediate response with unprecedented consequences. Three days later, he ordered Russian deterrence forces to be put on high combat alert as a form of nuclear signaling.[4]
- As Russia annexed four Ukrainian regions in September 2022, it escalated its nuclear rhetoric by claiming these areas were under its nuclear protection.[5] This was particularly concerning due to potential interpretations of Russia's military doctrine and policies. According to its Military Doctrine, nuclear weapons can be used in conventional conflicts "when the very existence of the state is threatened," suggesting that a significant war loss could be perceived as a threat to state survival, at least the survival of the regime.[6] Additionally, the Foundations of State Policy of Nuclear Deterrence highlight territorial integrity as a priority protected by Russia's nuclear deterrent, extending this protection to the newly occupied regions.[7]

On the one hand, a lot of the rhetoric was intended at deterring others from providing support to Ukraine. On the other hand, from the overall perspective, compellence is

---

[2]Thomas Shelling, *Arms and Influence* (London: Yale University Press, 2020), pp. 2–4; Polina Sinovets, *Double Faced Yanus or Nuclear Deterrence Theory in the XXI Century* (Odesa: Fenix, 2008), pp. 8–9.

[3]"Nuclear Signaling during the War in Ukraine." CSIS. Accessed May 9, 2024. https://nuclearrussiaukraine.csis.org/#about

[4]Telegraph, "Putin Declares Military Offensive in Ukraine as Invasion Starts," YouTube, February 24, 2022, https://youtu.be/_5YeX8eCLgA; Andrew Roth et al., "Putin Signals Escalation As He Puts Russia's Nuclear Force on High Alert," *Guardian*, February 28, 2023, https://www.theguardian.com/world/2022/feb/27/vladimir-putin-puts-russia-nuclear-deterrence-forces-on-high-alert-ukraine.

[5]President of Russia, "Address by the President of the Russian Federation," September 21, 2022, http://en.kremlin.ru/events/president/news/69390

[6]"Military Doctrine of the Russian Federation," Rossiyskaya Gazeta, December 30, 2014, https://rg.ru/documents/2014/12/30/doktrina-dok.html

[7]President of Russia, "Decree of the President of the Russian Federation of June 2, 2020, No. 355," http://www.kremlin.ru/acts/bank/45562

in attempting to change the status quo on the terms favorable to Russia and using nuclear status to shield a war of aggression.

Russia's control of the Zaporizhzhia Nuclear Power Plant (ZNPP) allowed leveraging the fear of nuclear incidents. While attacks on nuclear power plants are not considered the same as nuclear-weapon threats, the mere presence of "nuclear" can evoke similar apprehension—radiophobia. Radiophobia, the intense fear of ionizing radiation, has been a focus of study even before the atomic bombings of Hiroshima and Nagasaki in 1945, and has been heightened by subsequent disasters such as Chornobyl in 1986 and Fukushima in 2011.[8] Therefore, by seizing control of ZNPP, Russia gained the ability to employ hybrid compellence, leveraging the crucial factor of the opponents' threat perception.

In the process of occupying the facility, Russian forces shelled the ZNPP, hitting its training center near the reactor and sparking a fire on March 4, 2022.[9] Safe operation of an NPP depends largely on the functionality of all systems free from physical damage due to military action as well as safety, professionalism, and agility of individual operators. On the contrary, since the takeover, ZNPP faced significant challenges, including air strikes, shelling, a shortage of spare parts, and disruptions to the supply chain, frequent power outages, and deviations from planned activities and normal operations.[10] Reduced staffing levels, difficult working conditions, and even abductions, beatings, and coercion of personnel presented significantly elevated risks of another manmade radiological emergency.[11]

Reports indicate over 500 Russian troops were stationed at the plant, with heavy artillery and anti-personnel mines deployed along the Kakhov Reservoir shores.[12] This occupation allowed Russia to conduct military operations against Ukrainian forces across the reservoir, knowing retaliation would be difficult. Russia's actions signified the weaponization of critical infrastructure, risking a nuclear catastrophe to achieve military objectives.[13] This behavior mirrors the "game of chicken" in

[8] John C.H. Lindberg and Denali Archer, "Radiophobia: Useful Concept, or Ostracising Term?," *Progress in Nuclear Energy*, Vol. 149 (July 2022), DOI:10.1016/j.pnucene.2022.104280.

[9] *Nuclear Safety, Security and Safeguards in Ukraine.* (Vienna: IAEA, 2023), p.1, https://www.iaea.org/sites/default/files/23/02/nuclear-safety-security-and-safeguards-in-ukraine-feb-2023.pdf

[10] *Nuclear Safety, Security and Safeguards in Ukraine.* IAEA, p.5.

[11] Joe Parkinson and Drew Hinshaw, "'The Hole': Gruesome Accounts of Russian Occupation Emerge from Ukrainian Nuclear Plant," *Wall Street Journal*, November 18, 2022, https://www.wsj.com/articles/russian-occupation-of-nuclear-plant-turns-brutal-with-accusations-of-torture-and-beatings-11668786893?mod=hp_lead_pos3; Olena Badiuk, "'Nuclear Prison': how Russia Turned the ZNPP into a Large Torture Chamber by Involving Rosatom," *Radio Svoboda*, September 19, 2023, https://www.radiosvoboda.org/a/novyny-pryazovya-rosatom-sanktsiyi-zaporizka-aes-kativnya/32597882.html

[12] Drew Hinshaw and Joe Pakinson, "Russian Army Turns Ukraine's Largest Nuclear Plant into a Military Base," *Wall Street Journal*, July 5, 2022, https://www.wsj.com/articles/russian-army-turns-ukraines-largest-nuclear-plant-into-a-military-base-11657035694

[13] Valeriia Hesse, "How Doing Justice to Ukraine Can Help Win the Battle for the Future of Nuclear Security," Stimson Center, October 3, 2023, https://www.stimson.org/2023/how-doing-justice-to-ukraine-can-help-win-the-battle-for-the-future-of-nuclear-security/

nuclear coercion, where perceived irrationality and willingness to risk nuclear disaster can force opponents to back down.[14] This strategy exploited the sociopsychological impacts for coercive purposes. Thus, Russia's hybrid nuclear compellence manifested itself through both the use of nuclear weapons threats and threats posed to the civilian nuclear infrastructure.

3 Consequences

Russia's seeming irrationality effectively deterred other nations from intervening in the conflict and providing certain armaments to Ukraine. Even amidst debates about whether the rhetoric represented nuclear signaling or nuclear threats, the main point is that the nature of nuclear coercion assigns the importance to both intent and perception. The prevailing perception was that nuclear use could not be discounted.[15]

However, attempts at compellence against Ukrainians did not yield the anticipated results. Ukrainians took the threats seriously but stood firm, whereby the public provided widespread support for the official government stance.[16] The potential outcome from a Ukrainian capitulation far outweighed the fear of enduring a nuclear strike. Especially when the evidence suggests that nation as a whole would survive the aftermath of a nuclear disaster, as seen in modern-day Hiroshima and Nagasaki, and even through Ukraine's own experience of Chornobyl. Russia's genocidal rhetoric, the harrowing experiences in Bucha and other Russian-occupied territories, and the enduring historical trauma fueled continued resistance, which was perceived as a battle against annihilation.[17]

[14] Valeriia Hesse, "Winning the Game of Chicken With Memes: Ukrainian Reactions to Russian Threats," *Arms Control Today, June 2023,* https://www.armscontrol.org/act/2023-06/features/winning-game-chicken-memes-ukrainian-reactions-russian-threats#endnote06

[15] "Will Putin use nuclear weapons in Ukraine? Our experts answer three burning questions," Atlantic Council, May 10, 2022, https://www.atlanticcouncil.org/blogs/new-atlanticist/will-putin-use-nuclear-weapons-in-ukraine-our-experts-answer-three-burning-questions/

[16] Hesse, "Winning the Game of Chicken With Memes: Ukrainian Reactions to Russian Threats."

[17] Olesya Kotubeii, "Ethnic Cleansing: The Article 'What Russia Should Do With Ukraine' Was Published on the Russian State Website," Suspilne, April 4, 2022, https://suspilne.media/225052-etnicni-cistki-na-derzavnomu-rosijskomu-sajti-vijsla-statta-so-rosia-mae-zrobiti-z-ukrainou/; Richard Connor, "Russia's Ukraine Violations 'Shockingly Routine,'" Deutsche Welle, March 31, 2023, https://www.dw.com/en/russias-ukraine-violations-shockingly-routine/a-65196353; Roberto Cancio, Anastasiia Kuptsevych-Timmer, and Marisa Omori, "Perpetual War With the Brother Nation: An Analysis of Ukrainian Veterans, Cultural Identity and Historical Trauma," *Journal of War & Culture Studies*, Vol 13, No. 3 (2020): 219–236; Yaroslava Muzychenko, "The Rake of Russian Atrocities: How Putin's Russia Continues Three Hundred Years of Terror Against Ukraine," Radio Liberty, April 7, 2023, https://www.radiosvoboda.org/a/rosiya-zvirstva-viyna-ukrayina/31789267.html#comments

Remarkably, fear levels remained relatively low. In October 2022, only 35% of Ukrainians believed in the possibility of Russia employing nuclear weapons against Ukraine, with a mere 16% expecting a nuclear strike against NATO.[18] Furthermore, in 2024, Ukraine ranked among the countries least fearful of nuclear threats, as per polling data.[19]

Robust international diplomatic pressure bolstered by the Ukrainian public reaction in response to nuclear threats noticeably softened Russia's rhetoric, as it failed to achieve its intended impact. This can be observed in several prominent public comments by leading Russian figures: Putin's Valdai speech on October 27, 2022, subtly deflected statements regarding Russia's potential use of nuclear weapons; on November 2, 2022, the Russian Ministry of Foreign Affairs issued a statement emphasizing efforts to prevent nuclear conflict; and even Dmitriy Medvedev, in response to intensified Ukrainian strikes on Crimea, just promised retribution against "key figures of the Nazi regime."[20]

Similarly, the potential for a war-induced radiological emergency could not be dismissed. Military activities near the plant sparked fears among both the Ukrainian population and neighboring countries.[21] Early on, global concerns focused on the possibility of an environmental disaster, particularly if a fire broke out, dispersing radioactive particles across Europe with wind.[22]

In response to Russia's use of the ZNPP for hybrid compellence, international action options discussed publicly included similarly hybrid rhetoric. Thereby, a potential nuclear use and a potential attack on a nuclear plant with subsequent radiological emergency appeared under the same umbrella in the conversations.[23] The US Senate even introduced a resolution suggesting that actions by Russia, Belarus, or their proxies causing radioactive contamination on NATO territory

[18] "Eighteenth General National Survey: Attitudes of Ukrainians Towards Foreign Countries (October 8–9, 2022)," Rating, October 10, 2022, https://ratinggroup.ua/research/ukraine/v_s_mnadcyate_zagalnonac_onalne_opituvannya_stavlennya_ukra_nc_v_do_nozemnih_kra_n_8-9_zhovtnya_2022.html

[19] Iryna Balachuk, "Ukraine among countries where least fear nuclear threat, poll finds," *Ukrainska Pravda*, April 17, 2024, https://www.pravda.com.ua/eng/news/2024/04/17/7451628/

[20] Dmitry Medvedev, "The Kyiv Dog Continues to Bark," Telegram, April 29, 2023, https://t.me/medvedev_telegram/317

[21] Danica Kirka, "Russia's nuke plant attack revives Chernobyl disaster fears," AP, March 4, 2022, https://apnews.com/article/russia-ukraine-boris-johnson-science-business-europe-2167e4687c2f80bc0383feb266d8b085

[22] Valerie Hopkins and William J. Broad, "Combat at Ukraine Nuclear Plant Adds Radioactive Dangers to Russian Invasion," *The New York Times*, March 4, 2022, https://www.nytimes.com/2022/03/04/science/ukraine-nuclear-power-plant.html

[23] Natalie Musumeci and John Haltiwanger, "A Russian nuclear strike in Ukraine would 'almost certainly' trigger a military response from Kyiv's partners: senior NATO official," *Business Insider*, Oct 12, 2022, https://www.businessinsider.com/russian-nuclear-attack-almost-certainly-trigger-military-response-nato-official-2022-10

should be seen as an attack on NATO.[24] However, experts's assessments and models generally reassured Europeans that even in the worst-case scenario, the concentration of radiation would be too low to cause health issues.[25]

The international community and world leaders have been consistently calling for maintaining nuclear safety and preventing any man-made nuclear incidents at the ZNPP. These were best expressed through the International Atomic Energy Agency (IAEA) Director General Rafael Grossi's five **concrete principles to help ensure nuclear safety and security**:

1. No attacks of any kind should be directed at or from the plant, especially targeting reactors, spent fuel storage, critical infrastructure, or personnel.
2. ZNPP should not be used as a storage site or base for heavy weapons (e.g., multiple rocket launchers, artillery systems, munitions, and tanks) or military personnel that could facilitate an attack from the plant.
3. The plant's off-site power should remain secure. Efforts must ensure its availability and security at all times.
4. All structures, systems, and components vital for the safe and secure operation of ZNPP must be protected from attacks or sabotage.
5. No actions should undermine these principles.[26]

According to Article 51 of the UN Charter, Ukraine has the right to reclaim its territory.[27] The state attempted to retake the plant at least three times while adhering to the aforementioned principles, initially articulated through the Seven Indispensable Pillars for ensuring nuclear safety and security during armed conflict.[28] In October 2022, an infantry-only amphibious operation by Ukraine was launched under the assumption that Russian forces would not use artillery against troops

---

[24] Georgi Gotev, "Nuclear cloud will trigger NATO's Article 5, US warns Russia," *EURACTIV*, June 24, 2023, https://www.euractiv.com/section/global-europe/news/nuclear-cloud-will-trigger-natos-article-5-us-warns-russia/

[25] "Attack on Ukrainian Nuclear Power Plants: People in Hungary Have No Reason to Fear Disaster," *Hungary Today*, March 12, 2022, https://hungarytoday.hu/ukrain-nuclear-plants-effect-hungary-aszodi/; "Ukraine warns of Russia nuclear plant attack plot," *DW*, June 22, 2023, https://www.dw.com/en/ukraine-warns-of-russia-nuclear-plant-attack-plot/a-66002826

[26] "Is Ukraine's Zaporizhzhia nuclear plant at risk of an 'accident'?," *Al Jazeera*, April 16, 2024, https://www.aljazeera.com/news/2024/4/16/is-ukraines-zaphorizhzhia-nuclear-plant-at-risk-of-an-accident; Khrystyna Bondarieva, "EU top diplomat says Russia should withdraw from Zaporizhzhia Nuclear Power Plant," *European Pravda*, April 9, 2024, https://www.eurointegration.com.ua/eng/news/2024/04/9/7183403/; "IAEA Director General Statement to United Nations Security Council," IAEA, May 30, 2023, https://www.iaea.org/newscenter/statements/iaea-director-general-statement-to-united-nations-security-council-30-may-2023

[27] "Article 51—Right of Individual or Collective Self-Defense," United Nations, https://www.un.org/en/about-us/un-charter/full-text#:~:text=Article%2051,maintain%20international%20peace%20and%20security

[28] *Nuclear Safety, Security and Safeguards in Ukraine.* IAEA, p.6.

near a nuclear plant due to severe risks.[29] However, Russian forces began shelling the Ukrainian troops using tanks and artillery as they approached via the Kakhovka Reservoir, forcing a retreat to avoid nuclear disaster.[30] With the recent increase in military activity in the surrounding areas the IAEA itself has confirmed regular occurrence of outgoing artillery fire in multiple IAEA Director General Statements on Situation in Ukraine since March 2024.[31]

With impending Ukrainian counteroffensive in the region in summer 2023, Russia kept claiming that Ukrainian shelling around the plant could lead to a radioactive incident thus trying to discourage Ukraine from conducting its operations and pressure its allies to push against such a counteroffensive. Analysts had suggested that Russia might be preparing a false flag operation to make it appear as though Ukraine was responsible.[32] The current escalation, involving off-the-shelf drones, complicates the situation by giving the perpetrator plausible deniability.[33]

Russia's tactics and strategy pose a dangerous precedent, potentially encouraging similar abuses in future conflicts. Ukraine, striving to act as a responsible international player and garner support, finds itself disadvantaged by this hybrid compellence. Although nuclear power plants (NPPs) and nuclear weapons (NWs) are not equivalent, this novel weaponization method has sparked fears and discussions about potential future implications.

There is no comprehensive, legally binding international document specifically addressing attacks on nuclear facilities. At the same time, the Treaty on the Non-Proliferation of Nuclear Weapons (NPT) addresses non-compliance only when a non-nuclear weapons state violates the IAEA safeguards agreement. It lacks specific mechanisms to handle non-compliance by nuclear weapons states

[29] Maryna Lisnychuk, "The Armed Forces of Ukraine Tried to Liberate the ZNNP from the Occupiers in October, but Were Forced to Retreat for a Reason," *Obozrevatel*, April 8, 2023, https://war.obozrevatel.com/ukr/zsu-namagalisya-zvilniti-vid-okupantiv-zaes-u-zhovtni-ale-musili-buli-vidstupiti-cherez-vagomu-prichinu-times.htm. François Diaz-Maurin, "A Ukrainian Expert Assesses the Possible Impact of a Military Attack on the Zaporizhzhia Nuclear Power Plant," *Bulletin of Atomic Scientists*, September 2, 2022, https://thebulletin.org/2022/09/a-ukrainian-expert-assesses-the-possible-impact-of-a-military-attack-on-the-zaporizhzhia-nuclear-power-plant/

[30] Maryna Lisnychuk, "The Armed Forces of Ukraine Forced to Retreat for a Reason."

[31] "Update 224—IAEA Director General Statement on Situation in Ukraine," IAEA, April 18, 2024, https://www.iaea.org/newscenter/pressreleases/update-224-iaea-director-general-statement-on-situation-in-ukraine

[32] Christian Edwards, "Ukraine warns Russia might attack the Zaporizhzhia nuclear power plant. How worried should we be?," *CNN*, July 5, 2023, https://edition.cnn.com/2023/07/05/europe/zaporizhzhia-nuclear-disaster-threat-explainer-intl/index.html

[33] "Ukraine: Briefing on the Zaporizhzhia Nuclear Power Plant," Security Council Report, April 12, 2024, https://www.securitycouncilreport.org/whatsinblue/2024/04/ukraine-briefing-on-the-zaporizhzhia-nuclear-power-plant-3.php; drone photo: https://ria.ru/20240408/zaes-1938689080.html

with their security assurances and or by them using offensive compellence.[34] However, since the NPT unites both nuclear weapons and peaceful nuclear infrastructure under its pillars, one of the ways to address these new challenges could be to address them in the NPT Review Process. Despite encountering numerous challenges in recent decades, it is a trusted and actively functioning forum unlike, for example, Conference on Disarmament. It is almost certain that such a negotiation will be tough, controversial, and slow-progressing. However, a crucial principle could be the systematic incorporation of both convergent and divergent perspectives among state parties within the proposed outcome document in the event of a deadlock. As proposed in a recently published paper, the negotiation progress and language could be just reflected in an non-binding Information Circular.[35]

4 Conclusion

In the game of chicken manner, Russia managed to leverage nuclear fears in the form of hybrid compellence: threats involving the use of nuclear weapons and the manipulation of radiological emergency risks, both of which exploit public fear of radiation. The intention behind this is to change the status quo on Ukrainian territorial integrity in an offensive war and solidify its gains while simultaneously impeding external support to Ukraine. Such an approach made a significant input into external help being constantly delayed and Ukrainian counteroffensives being complicated. While Russia succeeded at coercing external powers with its nuclear weapons, under the current circumstances, the radiological emergency perspective does not frighten Ukrainian neighbors as much. Simultaneously, while nuclear weapons did not frighten Ukrainians to a point of giving in (they would be facing an existential threat even if they were to accept Russian terms); radiological emergency risks, which would directly affect its territory, and even more so international community's reaction to them, constrain Ukraine's actions. Thus, Russia's hybrid compellence had a limited effect in the war and keeps declining on many if not all fronts. However, it sets an extremely dangerous precedent, which might be learned and used by other maliciously-intended powers. Therefore, this behavior must be condemned and addressed properly in a comprehensive legally binding international document, and the best forum for it is the NPT framework since it has both nuclear weapons and peaceful nuclear infrastructure under its umbrella.

[34] Valeriia Hesse, "U-Factor: Russia's War on Ukraine and the Deterrence vs. Disarmament Discussion. Pragmatic Internationalism." In: Vicente, Aderito, Sinovets, Polina, and Julien Theron (eds). *Russia's War on Ukraine.* Springer, 2023. https://doi.org/10.1007/978-3-031-32221-1_12

[35] Ian Fleming Zhou, Valeriia Hesse, Anna-Elisabeth Schmitz and Karina Touzinsky, "Achieving Success Beyond Final Documents: Recommendations for the 2026 NPT Review Conference," *Arms Control Today, May 2024,* https://www.armscontrol.org/act/2024-05/features/achieving-success-beyond-final-documents-recommendations-2026-npt-review

References

1. Namestnik V (2023, February 27) How Russian Propaganda Uses Intimidation Tactics. Detector Media. https://disinfo.detector.media/post/yak-rosiiska-propahanda-vykorystovuie-taktyku-zaliakuvannia
2. Kotiv O (2023, March 1) 'There is no chance of survival': how Ukraine Met Spring, and Putin's Winter Terror Failed. 24 Channel. https://24tv.ua/zalyakuvannya-ukrayini-zimoyu-plan-putina-zamoroziti-ukrayintsiv_n2265364
3. Mills C (2023) Russia's use of nuclear *Thre*ats during the Ukraine conflict. House of Commons Library, London. https://researchbriefings.files.parliament.uk/documents/CBP-9825/CBP-9825.pdf
4. Shelling T (2020) Arms and influence. Yale University Press, London
5. Sinovets P (2008) Double faced Yanus or nuclear deterrence theory in the XXI century. Fenix, Odesa
6. Nuclear Signaling during the War in Ukraine (n.d.) CSIS. https://nuclearrussiaukraine.csis.org/#about
7. Telegraph (2022, February 24) Putin Declares Military Offensive in Ukraine as Invasion Starts. YouTube. https://youtu.be/_5YeX8eCLgA
8. Roth A et al (2023, February 28) Putin signals escalation as he puts Russia's nuclear force on high alert. Guardian. https://www.theguardian.com/world/2022/feb/27/vladimir-putin-puts-russia-nuclear-deterrence-forces-on-high-alert-ukraine
9. President of Russia (2022, September 21) Address by the President of the Russian Federation. http://en.kremlin.ru/events/president/news/69390
10. Military Doctrine of the Russian Federation (2014, December 30) Rossiyskaya Gazeta. https://rg.ru/documents/2014/12/30/doktrina-dok.html
11. President of Russia (2020) Decree of the President of the Russian Federation of June 2, 2020, No. 355. http://www.kremlin.ru/acts/bank/45562
12. Lindberg JCH, Archer D (2022) Radiophobia: useful concept, or ostracising term? Prog Nucl Energy 149:104280. https://doi.org/10.1016/j.pnucene.2022.104280
13. IAEA (2023) Nuclear Safety, Security and Safeguards in Ukraine. IAEA, Vienna. https://www.iaea.org/sites/default/files/23/02/nuclear-safety-security-and-safeguards-in-ukraine-feb-2023.pdf
14. Parkinson J, Hinshaw D (2022) 'The Hole': gruesome accounts of Russian occupation emerge from Ukrainian nuclear plant. Wall Street Journal. https://www.wsj.com/articles/russian-occupation-of-nuclear-plant-turns-brutal-with-accusations-of-torture-and-beatings-11668786893?mod=hp_lead_pos3
15. Badiuk O (2023, September 19) 'Nuclear Prison': how Russia Turned the ZNPP into a Large Torture Chamber by Involving Rosatom. Radio Svoboda. https://www.radiosvoboda.org/a/novyny-pryazovya-rosatom-sanktsiyi-zaporizka-aes-kativnya/32597882.html
16. Hinshaw D, Pakinson J (2022, July 5) Russian Army Turns Ukraine's Largest Nuclear Plant into a Military Base. Wall Street Journal. https://www.wsj.com/articles/russian-army-turns-ukraines-largest-nuclear-plant-into-a-military-base-11657035694
17. Hesse V (2023, October 3). How doing justice to Ukraine can help win the battle for the future of nuclear security. Stimson Center. https://www.stimson.org/2023/how-doing-justice-to-ukraine-can-help-win-the-battle-for-the-future-of-nuclear-security/
18. Hesse V (2023) Winning the game of chicken with memes: Ukrainian reactions to Russian threats. Arms Control Today. https://www.armscontrol.org/act/2023-06/features/winning-game-chicken-memes-ukrainian-reactions-russian-threats#endnote06
19. Atlantic Council (2022, May 10) Will Putin use nuclear weapons in Ukraine? Our experts answer three burning questions. Atlantic Council. https://www.atlanticcouncil.org/blogs/new-atlanticist/will-putin-use-nuclear-weapons-in-ukraine-our-experts-answer-three-burning-questions/

20. Kotubeii O (2022, April 4) Ethnic cleansing: the article 'What Russia should do with Ukraine' was published on the Russian State Website. Suspilne. https://suspilne.media/225052-etnicni-cistki-na-derzavnomu-rosijskomu-sajti-vijsla-statta-so-rosia-mae-zrobiti-z-ukrainou/
21. Connor R (2023, March 31) Russia's Ukraine violations 'shockingly routine.' Deutsche Welle. https://www.dw.com/en/russias-ukraine-violations-shockingly-routine/a-65196353
22. Cancio R, Kuptsevych-Timmer A, Omori M (2020) Perpetual war with the brother nation: an analysis of Ukrainian veterans, cultural identity and historical trauma. J War Cult Stud 13(3): 219–236
23. Muzychenko Y (2023, April 7) The Rake of Russian Atrocities: how Putin's Russia continues three hundred years of terror against Ukraine. Radio Liberty. https://www.radiosvoboda.org/a/rosiya-zvirstva-viyna-ukrayina/31789267.html#comments
24. Rating (2022, October 10) Eighteenth general national survey: attitudes of Ukrainians towards foreign countries (October 8–9, 2022)." Rating. https://ratinggroup.ua/research/ukraine/v_s_mnadcyate_zagalnonac_onalne_opituvannya_stavlennya_ukra_nc_v_do_nozemnih_kra_n_8-9_zhovtnya_2022.html
25. Balachuk I (2024, April 17) Ukraine among countries where least fear nuclear threat, poll finds. Ukrainska Pravda. https://www.pravda.com.ua/eng/news/2024/04/17/7451628/
26. Medvedev D (2023, April 29) The Kyiv Dog continues to bark. Telegram. https://t.me/medvedev_telegram/317
27. Kirka D (2022, March 4) Russia's nuke plant attack revives Chernobyl disaster fears. AP. March. https://apnews.com/article/russia-ukraine-boris-johnson-science-business-europe-2167e4687c2f80bc0383feb266d8b085
28. Hopkins V, Broad WJ (2022, March 4) Combat at Ukraine nuclear plant adds radioactive dangers to Russian invasion. The New York Times. https://www.nytimes.com/2022/03/04/science/ukraine-nuclear-power-plant.html
29. Musumeci N, Haltiwanger J (2022, October 12). A Russian nuclear strike in Ukraine would 'almost certainly' trigger a military response from Kyiv's partners: senior NATO official. Business Insider. https://www.businessinsider.com/russian-nuclear-attack-almost-certainly-trigger-military-response-nato-official-2022-10
30. Gotev G (2023, June 24) Nuclear cloud will trigger NATO's Article 5, US warns Russia. EURACTIV. https://www.euractiv.com/section/global-europe/news/nuclear-cloud-will-trigger-natos-article-5-us-warns-russia/
31. Hungary Today (2022, March 12) Attack on Ukrainian Nuclear power plants: people in Hungary have no reason to fear disaster. Hungary Today. https://hungarytoday.hu/ukrain-nuclear-plants-effect-hungary-aszodi/
32. DW (2023, June 22) Ukraine warns of Russia nuclear plant attack plot. DW. https://www.dw.com/en/ukraine-warns-of-russia-nuclear-plant-attack-plot/a-66002826
33. Al Jazeera (2024, April 16) Is Ukraine's Zaporizhzhia nuclear plant at risk of an 'accident'? Al Jazeera. https://www.aljazeera.com/news/2024/4/16/is-ukraines-zaphorizhzhia-nuclear-plant-at-risk-of-an-accident
34. Bondarieva K (2024, April 9) EU top diplomat says Russia should withdraw from Zaporizhzhia nuclear power plant. European Pravda. https://www.eurointegration.com.ua/eng/news/2024/04/9/7183403/
35. IAEA (2023, May 30) IAEA Director General Statement to United Nations Security Council. https://www.iaea.org/newscenter/statements/iaea-director-general-statement-to-united-nations-security-council-30-may-2023
36. United Nations (n.d.) Article 51 – right of individual or collective self-defence. United Nations. https://www.un.org/en/about-us/un-charter/full-text#:~:text=Article%2051,maintain%20international%20peace%20and%20security
37. Lisnychuk M (2023, April 8) The armed forces of Ukraine tried to liberate the ZNNP from the occupiers in October, but were forced to retreat for a reason. Obozrevatel. https://war.obozrevatel.com/ukr/zsu-namagalisya-zvilniti-vid-okupantiv-zaes-u-zhovtni-ale-musili-buli-vidstupiti-cherez-vagomu-prichinu-times.htm

38. Diaz-Maurin, F (222, September 2) A Ukrainian expert assesses the possible impact of a military attack on the Zaporizhzhia nuclear power plant. Bulletin of the Atomic Scientists. https://thebulletin.org/2022/09/a-ukrainian-expert-assesses-the-possible-impact-of-a-military-attack-on-the-zaporizhzhia-nuclear-power-plant/
39. IAEA (2024, April 18) Update 224 – IAEA Director General Statement on Situation in Ukraine. IAEA. https://www.iaea.org/newscenter/pressreleases/update-224-iaea-director-general-statement-on-situation-in-ukraine
40. Edwards C (2023, July 5) Ukraine warns Russia might attack the Zaporizhzhia nuclear power plant. How worried should we be? CNN. https://edition.cnn.com/2023/07/05/europe/zaporizhzhia-nuclear-disaster-threat-explainer-intl/index.html
41. Securitycouncilreport (2024, April 12) Ukraine: briefing on the Zaporizhzhia nuclear power plant. Security Council Report. https://www.securitycouncilreport.org/whatsinblue/2024/04/ukraine-briefing-on-the-zaporizhzhia-nuclear-power-plant-3.php
42. Hesse V (2023) U-factor: Russia's war on Ukraine and the deterrence vs. disarmament discussion. Pragmatic internationalism. In: Vicente A, Sinovets P, Theron J (eds) Russia's war on Ukraine. Springer. https://doi.org/10.1007/978-3-031-32221-1_12
43. Zhou I, Hesse V, Schmitz AA, Touzinsky K (2024, May) Achieving success beyond final documents: recommendations for the 2026 NPT review conference. Arms Control Today. https://www.armscontrol.org/act/2024-05/features/achieving-success-beyond-final-documents-recommendations-2026-npt-revie

